GROUND WATER MODELS
Scientific and Regulatory Applications

Water Science and Technology Board
Committee on Ground Water Modeling Assessment
Commission on Physical Sciences, Mathematics, and Resources
National Research Council

NATIONAL ACADEMY PRESS
Washington, D.C. 1990

National Academy Press • 2101 Constitution Avenue, N.W. • Washington, D.C. 20418

NOTICE: The project that is the subject of this report was approved by the Governing Board of the National Research Council, whose members are drawn from the councils of the National Academy of Sciences, the National Academy of Engineering, and the Institute of Medicine. The members of the committee responsible for the report were chosen for their special competences and with regard for appropriate balance.

This report has been reviewed by a group other than the authors according to procedures approved by a Report Review Committee consisting of members of the National Academy of Sciences, the National Academy of Engineering, and the Institute of Medicine.

The National Academy of Sciences is a private, nonprofit, self-perpetuating society of distinguished scholars engaged in scientific and engineering research, dedicated to the furtherance of science and technology and to their use for the general welfare. Upon the authority of the charter granted to it by the Congress in 1863, the Academy has a mandate that requires it to advise the federal government on scientific and technical matters. Dr. Frank Press is president of the National Academy of Sciences.

The National Academy of Engineering was established in 1964, under the charter of the National Academy of Sciences, as a parallel organization of outstanding engineers. It is autonomous in its administration and in the selection of its members, sharing with the National Academy of Sciences the responsibility for advising the federal government. The National Academy of Engineering also sponsors engineering programs aimed at meeting national needs, encourages education and research, and recognizes the superior achievements of engineers. Dr. Robert M. White is president of the National Academy of Engineering.

The Institute of Medicine was established in 1970 by the National Academy of Sciences to secure the services of eminent members of appropriate professions in the examination of policy matters pertaining to the health of the public. The Institute acts under the responsibility given to the National Academy of Sciences by its congressional charter to be an adviser to the federal government and, upon its own initiative, to identify issues of medical care, research, and education. Dr. Samuel O. Thier is president of the Institute of Medicine.

The National Research Council was organized by the National Academy of Sciences in 1916 to associate the broad community of science and technology with the Academy's purposes of furthering knowledge and advising the federal government. Functioning in accordance with general policies determined by the Academy, the Council has become the principal operating agency of both the National Academy of Sciences and the National Academy of Engineering in providing services to the government, the public, and the scientific and engineering communities. The Council is administered jointly by both Academies and the Institute of Medicine. Dr. Frank Press and Dr. Robert M. White are chairman and vice chairman, respectively, of the National Research Council.

Support for this project was provided by the Electric Power Research Institute under Contract No. RP2485-10, the U.S. Nuclear Regulatory Commission under Contract No. NRC-04-87-096, the U.S. Environmental Protection Agency under Contract No. CR-814067, the National Science Foundation under Grant No. CES/8708081, and the U.S. Army under Purchase Order No. DAAD05-88-M-M061.

Library of Congress Cataloging–in–Publication Data
Ground water models: scientific and regulatory applications /
Committee on Ground Water Modeling Assessment. Water Science and Technology Board, Commission on Physical Sciences, Mathematics, and Resources, National Research Council.
p. cm.
Includes bibliographical references.
ISBN 0-309-03993-2
1. Groundwater flow. 2. Liability for water pollution damages. I. National Research Council (U.S.). Committee on Ground Water Modeling Assessment.
TC176.G76 1989
363.73′94—dc20 89-14033
CIP

Printed in the United States of America

First Printing, January 1990
Second Printing, September 1990

COMMITTEE ON GROUND WATER MODELING ASSESSMENT

FRANK W. SCHWARTZ, Ohio State University, Columbus, *Chairman*
CHARLES B. ANDREWS, S. S. Papadopulos & Associates, Inc., Rockville, Maryland
DAVID L. FREYBERG, Stanford University, Stanford, California
CHARLES T. KINCAID, Battelle, Pacific Northwest Laboratory, Richland, Washington
LEONARD F. KONIKOW, U.S. Geological Survey, Reston, Virginia
CHESTER R. McKEE, In Situ, Inc., Laramie, Wyoming
DENNIS B. McLAUGHLIN, Massachusetts Institute of Technology, Cambridge
JAMES W. MERCER, GeoTrans, Inc., Herndon, Virginia
ELLEN J. QUINN, Northeast Utilities, Hartford, Connecticut
P. SURESH CHANDRA RAO, University of Florida, Gainesville
BRUCE E. RITTMANN, University of Illinois, Urbana-Champaign
DONALD D. RUNNELLS, University of Colorado, Boulder
PAUL K. M. van der HEIJDE, Holcomb Research Institute, Indianapolis, Indiana
WILLIAM J. WALSH, Pepper, Hamilton & Scheetz, Washington, D.C.

Staff

WENDY L. MELGIN, Project Manager
JEANNE AQUILINO, Project Secretary

Agency Liaisons to Committee

EDWARD H. BRYAN, National Science Foundation
DONALD L. CHERY, JR., U.S. Nuclear Regulatory Commission
STEPHEN R. CORDLE, U.S. Environmental Protection Agency
IRA MAY, U.S. Army Toxic and Hazardous Materials Agency
ISHWAR P. MURARKA, Electric Power Research Institute

WATER SCIENCE AND TECHNOLOGY BOARD

COMMISSION ON PHYSICAL SCIENCES, MATHEMATICS, AND RESOURCES

Preface

The issue of how ground water models should be used to address legal and regulatory concerns was brought to the attention of the Water Science and Technology Board in 1986. The U.S. Army requested assistance in assessing the efficacy of a specific modeling effort focused on an evaluation of contamination at a specific Army facility. The Army wanted to know to what extent that particular model could be used to apportion liability among several possible sources. The board concluded that investigation of such a site-specific problem was not appropriate for the NRC and decided instead to initiate a broader study dealing with the scientific basis and applicability of ground water models.

This initiative probably could not have come at a better time. Hydrogeologists are being caught in the middle between some major advances in science and increasing pressure from legal and regulatory bodies to use models to provide answers to specific questions. On the scientific side, there has been an explosion in knowledge in the past 10 years concerning the processes that control flow and mass transport in all kinds of hydrogeologic settings. This new understanding of how ground water systems behave has been incorporated in a variety of models. On the practical side, there is a community of users, employed by engineering consulting firms, government agencies, and national laboratories, who are being asked to solve increasingly complicated legal and regulatory problems. It is not at all clear whether existing models are appropriate for the tasks being set for them, nor is it clear to what extent new knowledge has changed modeling practice.

These issues provided the backdrop for an 18-month study supported by the Electric Power Research Institute, the U.S. Nuclear Regulatory Commission, the Environmental Protection Agency, the U.S. Army, and the National Science Foundation. The goal of this study was to address two questions: "To what extent can the current generation of ground water models accurately predict complex hydrogeologic and chemical phenomena?" and "Given the accuracy of these models, is it reasonable to assign liability for specific ground water contamination incidents to individual parties or make regulatory decisions based on long-term prediction?"

The members of the committee formed to address these topics came from universities, government, and private industry with broad experience related to the scientific and legal aspects of modeling. We benefited from the guidance and expert assistance provided by the staff of the Water Science and Technology Board. Our sponsors, to be sure, supplied money for the study, but, far more than that, they provided yet other points of view for us to consider and assistance in gathering information.

Our intention in writing this report was to produce a document that was stimulating, readable, and comprehensive. All of us came away from this task with some different views about models and modeling issues as the result of some very stimulating and provocative meetings. This report represents our best attempt at addressing some of the most controversial issues in ground water science today.

FRANK W. SCHWARTZ, Chairman
Committee on Ground Water
Modeling Assessment

Contents

Tables and Figures

TABLES

FIGURES

Overview, Conclusions, and Recommendations

OVERVIEW

Mathematical models, used commonly in ground water studies, are an attempt to represent processes by mathematical equations. The precise language of mathematics provides a powerful mechanism for expressing a tremendous quantity of information in an amazingly simple and compact way. Naturally, the starting point in modeling is a clear understanding of the processes involved. In terms of the flow of ground water or multiphase flow (i.e., when a fluid such as water, gasoline, or a dense nonaqueous-phase liquid is moving in the subsurface), one mainly needs to consider two dominant processes: flow in response to hydraulic potential gradients and the loss or gain of water from sinks or sources (e.g., pumping or injection, or gains and losses in storage). In the case of contaminant transport, a much larger number of diverse and complicated processes are involved. These processes can be divided into two groups: (1) those responsible for fluxes and (2) those responsible for sources and sinks for the material. Mass fluxes are prompted by processes like advection, diffusion, and mechanical dispersion. Sources and sinks are provided by a host of chemical, nuclear, and biological processes, such as sorption, ion exchange, oxidation/reduction, radioactive decay, and biodegradation.

In this report, Chapter 2 is devoted to explaining in a simple way how the differential equations for ground water flow and mass transport are created to embody the various processes. To fully describe a ground water system to be modeled, one needs in addition to the governing equations (1) specific numerical values for parameters that characterize the processes and for simulation parameters that are involved with the procedure to solve the equations and (2) information about the region, shape, and conditions along the boundaries. Solution of the resulting modeling problem is usually carried out analytically or numerically, depending upon the complexity of the hydrogeologic setting and the number of processes that need to be considered.

Few flow and transport problems can be modeled with confidence. As the following discussion explains, the most satisfactory results to date have come with models involving the flow of water or the transport of a single nonreactive contaminant in a saturated porous medium. As systems become more complicated through partial saturation, the presence of several mobile fluids, fracturing, or the existence of reacting contaminants, many more questions arise about the adequacy or validity of the underlying process models. The natural reaction of researchers is to undertake long-term experimental investigations, which in the scientific tradition will gradually improve our understanding of these processes. Although such research is undeniably important, it may not provide answers in time to influence many important national and local decisions about ground water contamination.

Chapters 3 and 4 of this report, along with many other thoughtful reports, papers, and articles, reveal major areas of uncertainty about subsurface contamination. Decisionmakers need to confront this uncertainty realistically and not be misled by the ability of computer models to always provide answers. Admitting the presence of uncertainty, however, is not enough. There is a need to make decisions, clean up water supplies, remove threats to public health, and devise safer methods for disposing of our wastes. Some of the decisions made in the short term may be inappropriate, inefficient, or even counterproductive, but it is unacceptable to simply wait until poorly understood environmental problems can be solved with more confidence.

In order to examine this issue further, it is useful to briefly review those areas where the understanding appears to be relatively good and those areas where there is still much to learn. Each of the major modeling categories discussed in Chapters 3 and 4 is briefly examined in the following paragraphs. Then the question of what decisionmakers can and should do now with problems requiring immediate attention is revisited.

The processes that control saturated ground water flow are reasonably well understood, and standard models of these processes are generally believed to be able to give reliable predictions if provided with adequate amounts of data. Nevertheless, the impacts of field-scale heterogeneity are still widely debated, and there are few clear guidelines on how model inputs should be estimated from limited data-bases or on how hydrologic monitoring programs should be designed. While saturated flow modeling is becoming more straightforward than it once was, there is much room for individual judgment, and the experience of the modeler still makes a significant difference in the quality of the results obtained. It is questionable whether this experience will ever be replaced by automated techniques such as expert systems, although such innovations may make the job of the informed modeler easier.

Unsaturated flow is less well understood. The basic "laws" that govern such flow are still questioned by some investigators. Much of the conventional theory of unsaturated flow is based on small-scale, one-dimensional laboratory experiments, which may not provide an accurate picture of behavior at larger field scales.

There have been very few field studies of unsaturated flow that extend over the scales of interest in most contamination applications, and most of these have focused on one-dimensional transport in the vertical direction. Some investigators believe that unsaturated flow can move horizontally over significant distances, although available evidence is insufficient to either confirm or reject this hypothesis.

Even if straightforward extrapolation from the laboratory to the field were possible, current techniques for determining unsaturated soil properties are too expensive and time-consuming to provide adequate descriptions of most contaminated sites. The numerical demands of all but the simplest unsaturated zone simulation models are formidable, and accurate three-dimensional unsaturated flow modeling capabilities are not available to most consultants. Yet many important contamination problems, such as leaking underground storage tanks, infiltrating pesticides, and leaching mining wastes, affect the unsaturated zone. Contaminant transport in this zone has only recently been perceived by the hydrologic community as an important research priority. Much remains to be done.

Flow through fractured media may be either saturated or unsaturated. Both types of fracture flow are difficult to predict at a given site unless extensive information is available about the fracture network. In this sense, true predictive modeling is not yet a reality. Nevertheless, recent research has provided significant advances in the understanding of the relative importance of the fracture and matrix systems in fractured flow. These advances have influenced some analyses of candidate radioactive waste disposal sites but have not, for the most part, reached the larger modeling community. The prevailing approach is to ignore fracture flow and hope that the effects of individual fractures will, in some sense, "average out." This can be a misleading oversimplification in some applications, where fractures can act as conduits for contaminant flow or can significantly modify subsurface flow patterns. Practical modelers need better guidelines for determining when fracture flow may be important and better methods for incorporating such flow into their model predictions.

The status of contaminant transport modeling depends greatly on the chemical species and phase of interest. In general, the processes that influence the transport of dilute, nonreactive aqueous phase solutes are well understood, at least in saturated media. There is, however, still widespread disagreement about the effects of spatial and temporal variability and about the related concept of macrodispersion. Until very recently, there were very few controlled field studies of ground water contaminant transport. Recent studies tend to indicate that real-world contaminant plumes have complex three-dimensional structures, which can be difficult to predict when soil properties are very heterogeneous. It can be difficult to simply map an existing plume, given the data typically available at a newly discovered contaminated site. Prediction of plume movement over many years is an even more difficult task.

The problems associated with transport modeling are greatly compounded when the solutes are reactive. In this case, chemical rather than hydrologic processes may govern the behavior of a contaminant plume. Ground water chemistry and ecology are relatively new fields that have had to contend with the problems inherent in working in an environment where processes are not readily observed and where samples are costly and difficult to obtain. Most models of reactive solutes are based on small-scale laboratory studies, which may not accurately mimic conditions found in the actual subsurface environment. This raises all of the same scale issues mentioned earlier in conjunction with unsaturated flow. Despite these difficulties, simple reactive transport models are in wide use and many modelers are aware of the need to at least consider sorption, biodegradation, and other chemical effects. It remains to be seen whether these simple models are adequate for decisionmaking purposes.

Most ground water contaminant modelers would probably agree that multiphase contaminant transport is the area where the basic physical mechanisms that control contaminant movement and degradation are least well understood and most difficult to model. Yet a wide range of important contaminants probably travel as separate liquid or gaseous phases when they move through the subsurface environment. Field-scale experimental investigations of multiphase transport are very limited, and existing laboratory-scale results indicate that this type of transport is influenced by a number of interacting factors, including viscosity and density contrasts, capillarity, and phase transitions. Although models of multiphase transport are available, many of the inputs they require are, as in the related case of unsaturated flow, difficult to estimate in a field setting. Because field data are very limited, it is practically impossible to confirm whether or not these models accurately reflect reality. Moreover, existing multiphase modeling techniques are computationally demanding and probably impractical to apply in situations where dozens of different interacting species and phases coexist. Such situations occur frequently. Leaking gasoline storage tanks are just one example.

Case studies provide a useful way to illustrate the application of models in (1) understanding ground water systems, (2) predicting contaminant migrations, and (3) decisionmaking by regulatory agencies. An example of the first type of application relates to the use of the generic vertical-horizontal spread (VHS) model by the U.S. Environmental Protection Agency (EPA) to determine when solid wastes need to be treated as hazardous wastes. In the case of the Madison aquifer, modeling studies predicted water-level declines due to large withdrawals by pumping. An example of the second type of application, modeling in connection with contamination of the Snake River plain, provided a prediction of the future extent of plume development. The third type of application is illustrated by the cases of contamination at the S-Area landfill in Niagara Falls and at Tucson Airport, where modeling was an integral part of the legal decisionmaking.

The above review of the present state of ground water contaminant modeling is not really as pessimistic as it may appear at first glance. In the last several years there has been substantial progress in such fields as fracture flow modeling, modeling of reactive transport and transformation, and multiphase modeling. The present concern with ground water contamination has stimulated a major increase in research efforts that has resulted in advancement. Moreover, it is the committee's opinion that the needs of decisionmakers are best served by honest and realistic assessments of the modeling state of the art. With such assessments we can set priorities, make difficult decisions, and understand how to deal with pressing short-term problems.

The fact that many of the models used in practice have not been validated to a significant extent provides an important source of uncertainty in the predictions that come from the models. Unfortunately, even more uncertainty enters the modeling process from, for example, (1) the inability to precisely describe the natural variability of model parameters (e.g., hydraulic conductivity) from a finite and usually small number of measurement points, (2) the inherent randomness of geologic and hydrogeologic processes (e.g., recharge rates and erosion) over the long term, (3) the inability to measure or otherwise quantify certain critical parameters (e.g., features of the geometry of fracture networks), and (4) biases or measurement errors that are part of common field methods. When all these sources of uncertainty are properly considered, a single model prediction realistically has to be viewed as one of a relatively large number of possible system responses. Over the past decade, the development of stochastic modeling techniques has been useful in quantitatively establishing the extent to which uncertainty in model input translates to uncertainty in model prediction.

To return to the question posed earlier, what should a decisionmaker do now, given existing modeling capabilities? There is obviously no easy or comforting answer to this question. It seems apparent, however, that it would be unwise to rely solely on any single source of information when deciding how to formulate regulations, carry out a cleanup, or protect public health. Models should be supplemented by carefully conceived field work, which not only provides data for estimating model inputs but also provides an independent confirmation of conditions in the subsurface environment. Put simply, the decisionmaker should hedge his bets and distribute his resources, funding different types of modeling efforts and mixing modeling with on-site monitoring. When field data are inconclusive or insufficient, model results may have a significant influence on the predicted impact of a given decision. In this case, the decisionmaker should request a quantitative and defensible assessment of the model's accuracy in order to evaluate the risk of making a bad decision. In this regard, environmental management is no different from any other form of management where uncertainty and risk are important. Models are not going to relieve us of the burden of making difficult decisions. They simply provide some additional information to consider. It is unrealistic to expect much more.

CONCLUSIONS AND RECOMMENDATIONS

Models and Subsurface Processes

Conclusions

There is a range of capability in modeling fluid flow in geologic media. Modeling saturated flow in porous media is straightforward with few conceptual or numerical problems. At the present time, conceptual issues and/or problems in obtaining data on parameter values limit the reliability and therefore the applicability of flow models involving unsaturated media, fractured media, or two or more liquids.

As a group, flow processes are among the most widely characterized hydrogeologic processes. The theories of flow involving either one or more fluids in porous and/or fractured media are well established and generally accepted. For the simplest cases involving saturated flow in porous media, the basic theoretical models have been validated in countless field and laboratory studies. The greatest source of uncertainty in prediction lies in supplying values of site-specific parameters. Flow in the unsaturated zone is less well understood, particularly in the case of dry soils, where the transport of water vapor can be significant. As was the case with saturated flow, establishing values for the controlling parameters under natural conditions is difficult, particularly for parameters like permeability that can vary in a complex, nonlinear way with moisture content.

Flow models involving two or more liquids in porous media are even more complicated in terms of the processes and parameters. Nevertheless, such models have been used and applied successfully, for example, in the petroleum industry. The greatest source of uncertainty in prediction remains the difficulty in accurately describing the spatial variability in controlling parameters. This problem of data is compounded by the variety of organic liquids that can be present as contaminants and for which specific experimental data are scarce.

In the case of fractured media, it remains to be shown through field and laboratory experiments that existing conceptual models of fractured systems are valid, particularly for cases involving variable saturation and more than one liquid. In addition, there are probably classes of fractured media that cannot be modeled with continuum theories and for which discrete approaches are impractical. The data problems remain. Many (controlling) parameters are difficult to measure or estimate accurately. Thus predictions for these more complex conditions need to be evaluated carefully and assessed in light of possible limitations.

Mass transport is controlled by a variety of physical, chemical, and biological processes. Quantitative descriptions of the processes concerned with mass transport (advection, diffusion, and dispersion) along with certain mass transfer processes (radioactive decay and sorption) are well understood. Multidimensional models of these processes have been used successfully in practice. Work is still required to account for other more complicated chemical processes (e.g., oxidation/reduction, precipitation, hydrolysis, and complexation) and biological processes (e.g., bacterial degradation) in mass transport models. Although prototype models exist for these more complicated systems, they are not yet developed for use in practice.

Contaminant transport is the outcome of mass transport processes, such as advection, diffusion, and mechanical dispersion, that move the mass and a multitude of mass transfer processes that redistribute mass within or between phases through chemical and biological reactions. Present-day understanding of mass transport developed from early studies on laboratory columns and more recent well-documented tracer studies in the field. The basic theory of advective transport modified by diffusion and mechanical dispersion is embodied in the familiar advection-dispersion equation, which provides a practical framework for modeling contaminant transport. The main source of uncertainty in prediction lies in establishing values of controlling parameters like velocity, effective diffusion coefficient, and dispersivity, which can be difficult to measure or estimate and vary spatially.

The complete description of mass transport usually requires that various chemical and/or biological processes also be considered. In the case of reactions such as radioactive decay, sorption, and hydrolysis, kinetic or equilibrium models describe the reactions and the necessary rate parameters or equilibrium constants for the reactions. These reactions can be calculated and measured with reasonable accuracy if not tabulated and can be incorporated in contaminant transport models in a straightforward manner. Although models for important reactions like oxidation/reduction, precipitation, and biodegradation exist, they are complicated to formulate and solve, difficult to characterize in terms of kinetic parameters, and largely unvalidated in practical applications. Thus the transport of multiple reacting constituents such as trace metals and organic compounds cannot be modeled with confidence.

As was the experience with flow, fracturing adds considerable complexity to mass transport. The issue of whether fractures are open or highly channelized, the importance of diffusion into the matrix, and how mixing occurs at fracture intersections make conceptualization of even mass transport processes uncertain. Coupled with the difficulty in formulating the model in terms of processes is the general lack of field and experimental data to validate models that are available. Thus transport modeling in fractured systems remains a highly speculative exercise.

Models and Decisionmaking

Conclusions

Properly applied models are useful tools to

- **assist in problem evaluation,**
- **design remedial strategy,**
- **conceptualize and study flow processes,**
- **provide additional information for decisionmaking, and**
- **recognize limitations in data and guide collection of new data.**

Ground water models are valuable tools that can be used to help understand the movement of water and chemicals in the subsurface. The purpose of the models is to simulate subsurface conditions and to allow prediction of chemical migration. When properly applied, models can supply useful information about flow and transport processes and can assist in the design of remedial programs.

The results of a model application are dependent on the quality of the data used as input for the model. Generally, site-specific data are required to develop a model of a site. The model cannot be used as a substitute for data collection. However, model use can help direct a data collection program by identifying areas where additional data are required. Closely linked data collection and model application can provide an adequate representation of site conditions. Incorrect model use frequently occurs when the limitations of the data used to develop the model are not recognized.

When properly applied, the results of a ground water model application can help in making decisions about site conditions. Model results can be used to supplement knowledge of site conditions but cannot be used to replace the decisionmaking process. The results of the models must be evaluated with other information about site conditions to make decisions about ground water development and cleanup.

Generic models are useful as a tool for initial screening but can never be used as a replacement for site-specific models.

Geologic materials are characteristically heterogeneous. The heterogeneity is seen at all scales, ranging from individual laminae a few millimeters thick to entire formations, aquifers, and drainage basins. In contrast, ground water models commonly incorporate various simplifying assumptions. Examples of some simplifications commonly used in ground water models include the assumption that an aquifer consists of a perfectly homogeneous, elastic material, or that the aquifer is made up of a small number of alternating homogeneous layers. The differences between the geologic reality of heterogeneity and the simplifications that may be used in ground water models make it scientifically dangerous and potentially misleading to blindly apply generic ground water models to any specific hydrogeological situation.

A generic model may be useful in offering some initial guidance to an investigator. However, only the most naive would rely on the predictions of a generic model in an attempt to understand the details of the movement of ground water or the behavior of a dissolved pollutant in a specific hydrogeological environment. It is essential that an investigator gather site-specific information to use as input to the ground water model of choice and, perhaps, that the model itself be modified and adapted to fit the hydrogeologic conditions at a particular site.

The results of mathematical computer models may appear more

certain than they really are; decisionmakers must be aware of the limitations.

Modelers must contend with the practical reality that model results, more than other expressions of professional judgment, have the capacity to appear more certain, more precise, and more authoritative than they really are. Many people who are using or relying upon the results of contaminant transport models are not fully aware of the assumptions and idealizations that are incorporated into them or of the limitations of the state of the art. There is a danger that some may infer from the smoothness of the computer graphics or the number of decimal places that appear on the tabulation of the calculations a level of accuracy that far exceeds that of the model. There are inherent inaccuracies in the theoretical equations, the boundary conditions, and other conditions and in the codes. Special care therefore must be taken in the presentation of modeling results. Modelers must understand the legal framework within which their work is used. Similarly, decisionmakers, whether they operate in agencies or in courts, must understand the limitations of models.

There are situations where government regulations require the use of contaminant transport modeling. As a general rule, however, it is not necessary for regulations to specify that a model must be used.

A few existing government regulations require that a model be used in the submission to the agency. All of the examples the committee found involved situations where the law required a long-term prediction of the migration potential of wastes. In such situations, there is no alternative but the use of contaminant transport models.

A regulation that requires contaminant transport modeling reflects an implicit decision to require a given level of detail and allow a given level of uncertainty. When regulations require the use of a model, however, they do not imply that the solution to the problem is susceptible to a "black-box" model application. Quite the contrary, in the cases examined, the regulations seem to require contaminant transport modeling in the most complicated site-specific problems.

Several agencies have guidelines that encourage the use of contaminant transport models. There are many different types of models, model applications, modeling objectives, and legal frameworks. Agencies cannot specify a list of government-approved models. A model that is appropriate for one problem may not be, and probably is not, applicable to another problem. Such a list also tends to stifle

innovation and use of newer models. On the one hand, government officials become reluctant to accept a nonapproved model. On the other hand, the regulated community may use an agency-approved model simply because the costs of getting governmental approval will be less. Such a list may also appear to be an "implied warranty" of the model accuracy and therefore lead to misuse of the models.

It is impossible to specify by a generally applicable regulation a contaminant transport model that would be scientifically valid in all applications and over the typical life of a regulation.

In some circumstances, it may be appropriate to specify the use of a particular contaminant transport model. For example, after reviewing site-specific data from a hazardous waste site, an agency or private company may determine that a particular model could be appropriate to apply at the site and such a model may be specified in a consent decree or permit for specific purposes. When a model is used in such circumstances, the consent decree, permit, or other legally enforceable procedure should require actual monitoring to confirm the modeling results and be flexible enough to allow the model to be updated and modified on the basis of new data and recent scientific developments.

Recommendations

Models used in regulatory or legal proceedings should be available for evaluation.

Models used in regulatory or legal proceedings are required to undergo public comment and review by those whose interest may be affected. The documentation associated with the model therefore must enable any reviewer to

- understand what was done;
- evaluate the quality of the model, considering issues such as the extent to which the equations describe the actual processes (i.e., model validation) and the steps taken to verify that the code correctly solves the governing equations and is fully operational (i.e., code verification);
- evaluate the application of the model to a particular site; and
- distinguish between the scientific and policy input.

A list of approved models should not be sanctioned by a regulatory agency. Agencies should not require that specific models be used for site-specific application by regulation, policy, or guidance. Instead,

positive attributes such as good quality assurance (QA) and documentation should be mandated, and government agencies should continue to support and provide resources for the development of ground water modeling codes.

The regulatory agencies should not develop a list of sanctioned ground water models. Models are used to evaluate a wide range of subsurface conditions for a variety of purposes. Models can be used to gain an understanding of flow and chemical transport, to evaluate remedial alternatives, and to determine data collection needs. The type of problem being evaluated and the level of understanding required should dictate the model selection.

A list of government-approved models would limit the choice of numerical codes available for problem solving. Development of a list of government-sanctioned codes could also inhibit model development and innovation. Because the process of model approval would probably be lengthy, approved models are likely to lag behind the available state of the art. As previously discussed, the quality of results is dependent on the quality of the data input and the knowledge of the models. Sanctioning of codes would not eliminate the need for proper model application and could develop a false sense of adequacy or accuracy for model users.

Instead of sanctioning particular models, regulatory agencies should provide detailed, consistent procedures for the proper development and application of models. Detailed specifications of positive aspects need to be developed but should include (1) good documentation of a code's characteristics, capabilities, and use; (2) verification of the program structure and coding, including mass balance results; (3) model validation, including a comparison of model results with independently derived laboratory or field data and possibly other computer codes; and (4) independent scientific and technical review.

The guidance must also be written to avoid being misconstrued as providing a list of "approved" models. The mere approving mention of a model in agency guidance may appear to inexperienced and untrained agency personnel as indicating that such models are "approved" or "sanctioned." Agency guidance therefore must stress that the descriptions do not sanction the use of any particular model. Instead the guidance should stress best modeling practices or principles, described above, and ensure that only experienced and properly trained personnel are involved in the development and review of such models.

Modeling should be considered to be only one of several possible methods of assessing liability in cases of ground water contamination. Models should not replace sound scientific and engineering judgment.

Contaminant transport models can provide one of several possible methods for identifying contaminant sources or apportioning liability. However, it would be rare for modeling alone to provide an unequivocal answer to the question of whether and to what degree a potential source is in fact a source. Ground water models must not be viewed as "black-box" tools that eliminate or lessen the need for common sense and good scientific judgment.

Similarly, while models may be useful tools in regulatory decisionmaking, they cannot substitute for the decisionmaking process. Such decisions are almost always based on a wide range of factors. Thus model results with attendant uncertainties should be considered along with all other information in order to make informed regulatory decisions.

Maintaining Scientific Integrity

Conclusions

Ground water models do and should vary in complexity. The complexity of the model used to analyze a specific site should be determined by the type of problem being analyzed. While more complex models increase the range of situations that can be described, increasing complexity requires more input data, requires a higher level and range of skill of the modelers, and may introduce greater uncertainty in the output if input data are not available or of sufficient quality to specify the parameters of the model.

Appropriate and successful models of ground water flow and transport can range from simple analytical solutions for one-dimensional flow in a homogeneous aquifer to highly complicated numerical codes designed to simulate multiphase transport of reactive species in heterogeneous, three-dimensional porous media. A useful model need not simulate all the physical, chemical, and biological processes that are acting in the subsurface. The model that is appropriate for analyzing a particular problem should be determined primarily by the objectives of the study. Unfortunately, there are no set rules for determining the appropriate level of complexity. The selection of an appropriate model and an appropriate level of detail and complexity is subjective and dependent on the judgment and experience of the analysts and on the level of prior information about the system of interest. Managers and other users of model results must be made aware that these trade-offs and judgments have been made and that they may affect the reliability of the model.

Models must be matched to the objectives of the study. Efforts should be made to avoid using models that are more complicated than necessary. Overly complicated models require information that cannot be obtained reliably from field measurements, which introduces unnecessary uncertainty into the modeling output, and overly complicated models require more time and money to operate, which wastes resources. Because there are no set rules for selecting an appropriate model, it is essential that agencies and companies employ qualified and well-trained personnel.

One of the key requirements in successfully applying flow or contaminant transport models is good-quality, site-specific data. Such data provide feasible bounds on the possible range of controlling parameters or boundary conditions, thereby minimizing the impact of data uncertainty as a major source of uncertainty associated with model predictions. In cases where particular model parameters are not or cannot be characterized, model prediction becomes much less certain because predicted variables like hydraulic potential or concentration could take on a much broader range of possible values.

A variety of factors can contribute to uncertainty in model predictions. One of the most important is the inability to characterize a site in terms of the boundary conditions or the key parameters describing important flow and transport processes. This uncertainty in data results for two basic reasons. First is the issue of the absolute number of data points providing information about a given parameter. Even a relatively large number of data points may not provide a basis for estimating parameter values at locations between them with total accuracy. As the number of data points decreases, this uncertainty attached to a parameter estimate increases to the point where one finally cannot describe the spatial variability in detail and has to resort to a simple estimate like a mean value for a given unit. A second issue with data is the inability in some cases to measure or even accurately estimate values for necessary parameters. This problem is most serious in fractured rocks for both single-phase and multiphase flow, and for mass transport processes involving certain kinetic processes (e.g., biodegradation, redox, and precipitation) whose rates can be extremely variable and site specific.

These two problems increase the likelihood that in many model studies there are some data that cannot be specified with accuracy. Sensitivity analyses provide one important way of establishing the extent to which uncertainty in a given parameter contributes to uncertainty in a prediction. Such analyses in many instances can provide the justification for carrying out additional field and laboratory studies.

In general, data collection and model application should not be viewed as sequential tasks but as tasks that should be performed interactively, complementing each other.

Good documentation of ground water models throughout the modeling process is necessary because of the complexities involved.

A hydrogeologic computer model may be very complex, running to thousands of lines of code. It may include hundreds of separate parameters and equations to model the movement of the water and the transport and fate of dissolved components. For these reasons it is essential that a model be accompanied by clear and thorough documentation, and that the documentation include a set of test problems that can be employed throughout the history of the model to verify that it continues to work properly. Adequate plans for testing and documenting a model should become part of any quality assurance program. Technical review should also be included in quality assurance plans to ensure that models have been adequately tested.

In addition to the inherent complexity, it is common for any given model to undergo repeated modifications and revisions, either by the author or by subsequent users. Unless a record is kept of the modifications that are made to the code, and unless the operational accuracy of the code is periodically tested and verified, serious doubts may develop about the validity and applicability of the code.

In addition to the original documentation, at least two types of information should accompany the code throughout its lifetime. First, changes in the structure of the model or of the database should be documented. The documentation may be in the form of a written record that is appended to the original documentation, or it may be included as comment lines within the noncompiled code. Second, an original set of test problems, including sample input and output, should accompany the code so that all users can periodically verify that the code is functioning properly, especially if changes are made. This periodic verification of the operation and output of the code becomes especially critical if the model is to be used as a part of a regulatory or legal action.

There is no valid reason to use a model that is unavailable for evaluation and testing by other qualified investigators. Similarly, new or revised models should be accompanied by sufficient documentation, history, and test problems to allow other qualified investigators to properly evaluate the model and to compare its output with that of other models.

As ground water model usage has increased, a shortage of qualified staff capable of appropriately applying models has been identified.

In order to avoid model misuse, it is important that the model user have the training and background to understand the many processes occurring in the subsurface. Experienced staff having this training and background are insufficient in terms of the number of sites where models could potentially be used.

Recommendations

All models must be documented so that the derivation of the model can be understood and the results can be reproduced by anyone seeking to use the model.

The documentation should include, at a minimum,

- a description of the underlying problem;
- a description of the fundamental equations that conceptualize the solution to the problem;
- a list of all assumptions used in the model and the rationale for their use;
- a description of the code used in the model;
- a verification of model codes against other solutions to the problem to verify the accuracy;
- an application of the model to a problem with a known solution, albeit perhaps a simpler problem, and a comparison of the results with the known results;
- a sensitivity analysis;
- the results of a quality assurance program;
- the validation of the model;
- a list of prior uses of the model, if any;
- a clear identification of the site-specific data used in the application of the model;

- a characterization of the level of precision, accuracy, and degree of uncertainty in the model results;
- a description of the statutory/policy criteria, if any, used to shape and select the assumptions and the acceptable level of precision, accuracy, and uncertainty; and
- any other information that is essential to understanding or being able to replicate the results.

All models must state quantitatively, to the extent possible, and if not quantitatively, then qualitatively, the degree and direction of uncertainty in the model results and the time frame over which the model's prediction can be considered acceptable.

This description of the uncertainties must be given at the beginning of the documentation of the model and wherever the conclusions of the models are used or discussed; e.g., in the conclusion of the modeler's report, in the briefing memorandum to an agency decisionmaker relying on the model, in whole or part, to make a regulatory decision, in the preamble to an agency regulation, and in expert testimony concerning the results of the model.

The policy assumptions used in the model must be explicitly listed, and the rationale for making each assumption must be described in the documentation and wherever the conclusions of the model are used or discussed; e.g., in the conclusion of the modeler's report, in the briefing memorandum to an agency decisionmaker relying on the model, in the preamble to an agency regulation, in press releases and statements to the public, in presentations to Congress, and in expert testimony concerning the results of the model. To avoid the misuse of ground water flow and transport models, agencies and companies should employ qualified and well-trained personnel.

Ground water flow and transport models are complex computer codes. To ensure that the input data are appropriate, and that the output results are properly utilized and interpreted, it is important to employ properly trained and qualified individuals. These personnel must be expert in both ground water science and its mathematical representation.

A certain fascination exists among technical personnel regarding the use of these powerful tools, and it is tempting to view them as "black boxes" that somehow produce easy and exact answers to previously difficult problems. This tendency may become even more pronounced as the interfaces between the codes and the users become more "user friendly." Indeed, it could be argued that the lack of a user-friendly interface may be a useful safety feature to help prevent inappropriate use of the models by nonqualified personnel.

If governmental agencies or private companies make the decision to use computer models in their work with ground water, it is essential that the personnel involved be adequately trained and fully aware of the limitations of the code. In order to use ground water models, an organization may have to hire new personnel or train existing personnel. It is not acceptable, however, to assign modeling projects to existing personnel who may simply be available for such tasks, without intensive and appropriate training.

The best procedure to ensure competency may be to designate one or more people as specialists in the modeling efforts within an organization. Such specialists would then have the responsibility to continually maintain and update their knowledge of the models being used and to make certain that others within the organization do not use the models inappropriately.

The problem of rapid turnover of personnel within government regulatory agencies must also be addressed. Pressures can be very great on regulatory personnel, without corresponding financial rewards. The record of high turnover rates within regulatory agencies, especially among younger technical employees, shows that the temptation to move into the private sector is very great. The decision to leave government service seems to be made about the time the individual achieves a relatively high level of competence and becomes known to various private companies. To overcome this high rate of attrition, some means of providing appropriate financial compensation must be found to properly recognize, reward, and retain highly skilled individuals. If salaries cannot be raised, it is essential that an active program of recruitment and training be maintained within the agency to ensure that an adequate, high level of competency always exists among the personnel involved in ground water modeling.

Research should be undertaken to provide the field and laboratory data necessary to validate flow and transport models.

Given that some types of models cannot be validated with existing, rather limited knowledge about some types of flow and mass transport processes, it is recommended that research be undertaken to fill in information gaps. The committee recognizes a need for well-controlled field and laboratory experiments involving flow and mass transport in fractured media, and multicomponent transport of chemically and biologically active contaminants. Such work is essential to establish how well existing mathematical concepts describe actual hydrogeological systems.

Recommendations for the Future

Governments, academic institutions, and private industry need to provide financial resources and substantially increase the pool of qualified personnel in the spectrum of fields essential to ground water modeling.

A severe shortage of qualified personnel exists in the areas of hydrogeology, ground water hydrology, and organic and aqueous geochemistry. Most of the new positions are with engineering and environmental consulting firms, and severe recruiting pressure exists among the firms, especially for experienced people.

If the challenges posed to our ground water environment by an ever-increasing population and continued industrialization are to be met, significant steps to increase the supply of trained ground water professionals must be taken. It is the strong recommendation of the committee that additional educational resources be committed to these fields as quickly as possible. The committee also recommends that government and private industry join in the effort to increase educational resources and opportunities for students entering the spectrum of fields related to ground water modeling. In addition to providing financial support, governmental agencies and private industry should further help in the education of ground water professionals by developing traineeships and industrial-associates programs to give students the opportunity to obtain practical experience in the field.

Government agencies and private industry should be aware of the need for and benefits of additional research. Research should be pursued in the following areas:

- **validation and further development of models involved with (1) ground water flow in unsaturated and fractured media, (2) multiphase flow in porous and fractured media, and (3) mass transport coupled with chemical reaction;**
- **role of bacteria in the transport and removal of contaminants;**
- **models in decisionmaking, including methods for identifying and presenting uncertainty and for establishing the reliability of model results;**

- **process characterization through well-controlled field and laboratory studies; and**
- **development of new approaches for parameter estimation and of new measurement techniques.**

Although many aspects of ground water modeling have major deficiencies in terms of scientific understanding and the availability of field-relevant databases, research in the five areas listed here offers especially great potential for yielding useful results. In the case of the first area—flow and transport in fractured and cavernous media and multiple-phase flow—the potential benefit is very high because these types of flow situations have a relatively widespread occurrence, have a strong impact on the movement of large masses of contaminants, and have not been adequately documented, resulting in an utter lack of any reliable databases. The second area—the role of bacteria in the transport and removal of contaminants—is critical because of the increasing recognition that bacteria are present in the subsurface, that most organic and some inorganic contaminants are biotransformed, and that bioremediation offers a potentially economical in situ cleanup technique. The third area—the role of modeling in decisionmaking, including legal and social interactions—must be understood if the courts, enforcement agencies, industries, and the affected public are to obtain the benefits of modeling. The last two areas—characterization through well-controlled field and laboratory studies and development of new approaches for parameter estimation and new measurement techniques—are essential if fate, transport, and remediation are to be measured in the subsurface, which is otherwise not easily accessible to human observation.

1

Introduction

This report addresses the use of ground water flow and contaminant transport modeling in the regulatory process. Its goals are to (1) examine the scientific bases upon which existing models are founded; (2) communicate the philosophies and approaches routinely used in the application of models to decisionmaking for regulatory purposes; and (3) provide guidelines concerning how models should be developed and applied in the regulatory process so that their utility and credibility are enhanced. This study is particularly timely because there are both increasing reliance on models and increasing uncertainty about the extent to which models can be and should be used.

Because the subsurface environment is not easily observed or accessible, models have become the tools employed to understand ground water systems and simulate and predict their behavior. Models are nothing more than mathematical representations of complex phenomena (McGarity, 1985). They are used to do the following:

- evaluate the understanding of physical processes in a quantitative way;
- identify the key issues needing further theoretical or field research;
- educate a nontechnical audience such as a government poli-

cymaker or the public, including a judge or a jury, by illustrating a phenomenon or concept;

- select optimal sampling locations and otherwise enhance field monitoring;
- simulate the past or future response of water levels to pumping, or the pattern of spreading of a plume of chemicals from a landfill, spill, leaking underground storage tank, or other source;
- design a ground water remedial program; or
- optimize efficiency in industrial processes, such as secondary and tertiary methods of recovering oil.

Both flow and transport models have been used in an equally wide variety of regulatory and legal contexts, such as the following:

- a federal or state environmental impact statement (EIS) to assess the potential impact of a particular project before it is implemented, e.g., the likelihood and severity of leakage of radioactive wastes from a long-term nuclear waste depository;
- an administrative record to support the technical standards required pursuant to federal or state regulations;
- an administrative record supporting a remedial action decision;
- an administrative record for a permit at a particular site; and
- evidence at a trial, e.g., to establish causation in a Superfund contribution action by one private party against another private party or to establish exposure in a personal injury action.

THE GROWTH IN THE USE OF MODELS

The growth in the use of models in the United States stems from a series of ever more stringent and comprehensive environmental statutes developed since the early 1970s. The most important statutes for the purposes of this report include the Comprehensive Environmental Response, Compensation, and Liability Act (CERCLA or "Superfund"), the Resource Conservation and Recovery Act (RCRA), the Safe Drinking Water Act (SDWA), and the National Environmental Policy Act (NEPA) (see Table 1.1). The galvanizing force for these statutes came from highly publicized pollution incidents, such as the relocation of residents from the vicinity of contamination sources at Love Canal in New York, and Times Beach, Missouri.

There is a very large number of potential sources of ground water contamination (see Table 1.1). Virtually all of these sources could

TABLE 1.1 Estimated Numbers of Contamination Problems That Need to Be Addressed Under Various Statutes

Type of Potential Ground Water Contamination Source	Number Nationwide (unless otherwise noted)
Superfund hazardous waste National Priority List (NPL) sites (currently on the NPL or proposed)[a]	951
Potential Superfund NPL sites that must be assessed preliminarily and inspected by 1989[b]	27,000
Superfund remedial investigations and feasibility studies at Superfund sites that must be commenced	
By 1989	275
By 1991[c]	650
Superfund remedial actions that must be commenced	
By 1989	175
By 1992[c]	375
RCRA hazardous waste facilities[d]	
Operating landfills	393
Closing landfills	1,095
Operating and closing incinerator and other treatment and storage facilities	3,338
Projected number of RCRA facility investigations	2,938
RCRA nonhazardous waste facilities (e.g., municipal and identical landfills)[e]	70,419 to 261,930
RCRA nonhazardous waste facilities with a high likelihood of containing hazardous wastes[e]	70,419
Mining waste sites[e]	22,339
Underground storage tanks[e]	10,820
Pesticide-contaminated sites[e]	3,920
Underground injection wells[f]	
Class I wells (hazardous waste injected below a U.S. drinking water supply)	533
Class II wells (secondary oil and gas production)	153,126
Class III wells (mining)	249
Class IV wells (hazardous waste injected into or above a drinking water supply—now essentially banned)	25
Class V wells (all other miscellaneous wells)	46,271
TOTAL	200,204
Estimated number of abandoned and unplugged oil and gas wells[g]	1,200,000
Sites with releases of radioactive materials[e]	1,502
Environmental impact statements per year[h] (1985) (it is estimated that between 15 and 40 percent of these documents may involve projects that require the use of ground water or contaminant flow modeling)	549
Surface impoundments[i]	
Industrial	25,749
Municipal	36,179
Agricultural	19,167
Mining	24,451

TABLE 1.1 *Continued*

Type of Potential Ground Water Contamination Source	Number Nationwide (unless otherwise noted)
Oil/gas brine pits	64,951
Other	5,748
TOTAL	176,245
Petroleum product pipelines miles (1976)[i] (carrying 10 billion barrels)	175,000
Liquid petroleum and nonhazardous waste underground storage tanks (as of 1984)[k]	2,500,000

[a] National Priorities List for Uncontrolled Hazardous Waste Sites, 52 Fed. Reg. 27,620, 27,621 (1987) (Final Rule).
[b] Section 116(d)(1) of CERCLA, 42 USCA § 9616(d)(1); and Surveys and Investigations Staff, House Committee on Appropriations, *Report on the Status of the Environmental Protection Agency's Superfund Program* 31 (March 1988) (hereinafter House Staff Report).
[c] Section 116(e)(1) of CERCLA, 42 USCA § 9616(e)(1).
[d] General Accounting Office, *Hazardous Waste: Corrective Action Cleanups Will Take Years To Complete* Table II.1, at 31 (1987) (GAO/RCED-88-48).
[e] General Accounting Office, *Superfund: Extent of Nation's Potential Hazardous Waste Problem Still Unknown* Table 2.1, at 14 (1987) (GAO/RCED-88-44).
[f] General Accounting Office Report to the Chairman, Environment, Energy, and Natural Resources Subcommittee, Committee on Government Operations, House of Representatives, *Hazardous Waste: Controls Over Injection Well Disposal Operations* Table 1.2, at 13 (1987) (GAO/RCED-87-170).
[g] EPA, *Report to Congress on the Management of Wastes and the Exploration, Development, and Production of Crude Oil, Natural Gas, and Geothermal Energy, Executive Summaries* 14 (December 1987).
[h] Council on Environmental Quality, *The Sixteenth Annual Report of the Council on Environmental Quality* Table 4-4, at 173 (1986).
[i] Geophysics Study Committee, Geophysics Research Forum, Commission on Physical Sciences, Mathematics, and Resources, National Research Council, *Groundwater Contamination* Table 1, at 4 (1984).
[j] Patrick, R., E. Ford, and J. Quarles, *Groundwater Contamination in the United States* 269 (2d ed. 1987).
[k] House Staff Report, supra note *b*, at 13; also see G. Lucero, Director of the Office of Waste Programs Enforcement, EPA, *Son of Superfund, Can the Program Meet Expectations*, Environmental Forum 5, 5–9 (March/April 1988).

There is a very large number of potential sources of ground water contamination (see Table 1.1). Virtually all of these sources could require the use of a contaminant transport model. The use of models is increasing at an accelerated rate. Guidance on the investigation of hazardous waste sites by federal agencies will encourage the use of contaminant flow models in the future (see Chapter 6).

Many of the responsibilities mandated by federal and state legislation cannot adequately be carried out without models. Yet, the majority of federal and state agencies have no overall strategy for developing, using, disseminating, and maintaining these valuable tools

(Office of Technology Assessment, 1982). As we will see throughout this report, the key scientific question affecting whether a model can be used is: How good are the predictions made by the model? There are undeniable scientific uncertainties inherent in model predictions, e.g. (National Research Council, 1988),

> [t]here is no model that will adequately describe *all* ground water quality problems because the assumptions and simplifications generally associated with models do not adequately mimic all the processes that influence the movement and behavior of the water and/or the chemicals of interest.

Legal issues can also determine whether a model is used properly. How good do the predictions need to be as a matter of law and/or policy?

It is within this context that the Water Science and Technology Board assembled the Committee on Ground Water Modeling Assessment to examine the current state of knowledge in ground water models and the role of contaminant transport in the regulatory arena. This 21-month study was supported by the Electric Power Research Institute, the U.S. Nuclear Regulatory Commission (USNRC), the U.S. Environmental Protection Agency (EPA), the National Science Foundation, and the U.S. Army.

The remainder of this report is divided into six parts. Chapter 2 describes how models are classified, the mathematical formulation and solution of the flow and mass transport equations, and the steps that are followed in code selection and model development.

Chapters 3 and 4 provide basic background information in the form of an overview of the important physical, chemical, and biological processes that provide the scientific framework for models. The intent of these chapters is to give the reader a clear appreciation of how water and contaminants move in flow systems and which parameters control their behavior.

Chapter 5 reviews the agency regulations and guidelines that require or give guidance on the use of modeling and provides five case studies. This chapter demonstrates how the concepts of modeling, developed in the previous chapters, have been applied to practical problems.

Chapter 6 reviews the USNRC and EPA experience in applying models and discusses other issues in the development and use of models. For example, quality assurance, the lack of qualified modelers, and the role of modeling in management are discussed.

Chapter 7 focuses on what the committee perceives to be the

emerging scientific, engineering, and policy trends as they relate to modeling. Issues examined in this chapter include linking geochemical and physical transport models, developing new modeling capabilities to handle complex processes, and the emerging new model approaches.

The committee attempts to bring together the varied concepts and ideas that were developed throughout the report in a way that will be useful to regulators and modelers alike. As the reader will discover, there are inherent limitations in what models can accomplish, but there are ways in which the developers and consumers of these models can enhance their usability.

REFERENCES

McGarity, T. 1985. The Role of Regulatory Analysis in Regulation Decision-Making, (background report for Recommendation 85-2 of the Administrative Conference). Published by Administrative Conference of the United States, p. 241.

National Research Council. 1988. Hazardous Waste Site Management: Water Quality Issues. Report on a colloquium sponsored by the Water Science and Technology Board. National Academy Press, Washington, D.C., p. 14.

Office of Technology Assessment. 1982. Use of Models for Water Resources Management, Planning, and Policy. U.S. Government Printing Office, Washington, D.C., pp. 9–10.

2

Modeling of Processes

INTRODUCTION

This chapter describes what models are and how they work. It begins by explaining the processes that control ground water flow and contaminant transport. To understand models, it is necessary to describe these processes by using certain mathematical equations that quantitatively describe flow and transport. The mathematical aspects of modeling are critical. The precise language of mathematics provides one of the best ways to integrate and express knowledge about natural processes. By developing an awareness of the natural processes, the mathematics should be understandable. Also, where the process is not well understood, this awareness provides an appreciation of the limits of the mathematics. Methods of solving the mathematical expressions are presented at the end of the chapter.

Subsurface movement—whether of water, contaminants, or heat—is affected by various processes. These processes can be related to three different modeling problems: ground water flow, multiphase flow (e.g., soil, water, and air; water and gasoline; or water and a dense nonaqueous-phase liquid (NAPL)), and the flow of contaminants dissolved in ground water.

Ground Water Flow

Of these three problems, ground water flow is the simplest to characterize and understand. In most cases, models need to consider only two ground water flow processes: flow in response to hydraulic potential gradients, and the loss or gain of water from sinks or sources, recharge, or pumping from wells. Hydraulic potential gradients simply represent the difference in energy levels of water and are generated because precipitation that is added to a ground water system at high elevations has more potential energy or hydraulic head than water added at a lower elevation (Figure 2.1). The result of these potential differences is that water moves from areas of high potential to areas of lower potential. As rainfall or other recharge keeps supplying water to the flow system, ground water continues to flow. On a cross section, it is possible to represent the spatial variability in hydraulic potential existing along a flow system by using what are called equipotential lines (see Figure 2.1). The equipotential lines are contours of hydraulic potential within some area of interest. In some simple situations, the direction of ground water flow is perpendicular to these equipotential lines, as shown in Figure 2.1.

The actual distribution of hydraulic head observed for an area depends mainly on two factors, how much and where water is added and removed, and the hydraulic conductivity distribution that exists in the subsurface. Consider a few examples. Figure 2.2 illustrates the hydraulic head distribution for two different water table configurations. The water table effectively represents the top boundary of the saturated ground water system, and its configuration reflects different recharge conditions. In both cases, the bottom and sides of the section are considered to be impermeable (no flow). With a linear water table and recharge mainly at the right end of the system, a relatively smooth regional flow system develops (see Figure 2.2a). The second water table, representing significant local areas of recharge and discharge at three locations, shows a much different flow pattern (see Figure 2.2b). Instead of a broad regional trend, several small, local flow systems have developed.

Ground water flow patterns also depend on the hydraulic conductivity distribution. Figure 2.3 compares the pattern of ground water flow along a cross section where all properties except the hydraulic conductivity for each layers are kept constant. Each of the two layers shown is defined in terms of a hydraulic conductivity in the horizontal direction (K_h) and in the vertical direction (K_v), with the ratio K_h/K_v describing the degree of directional dependence in

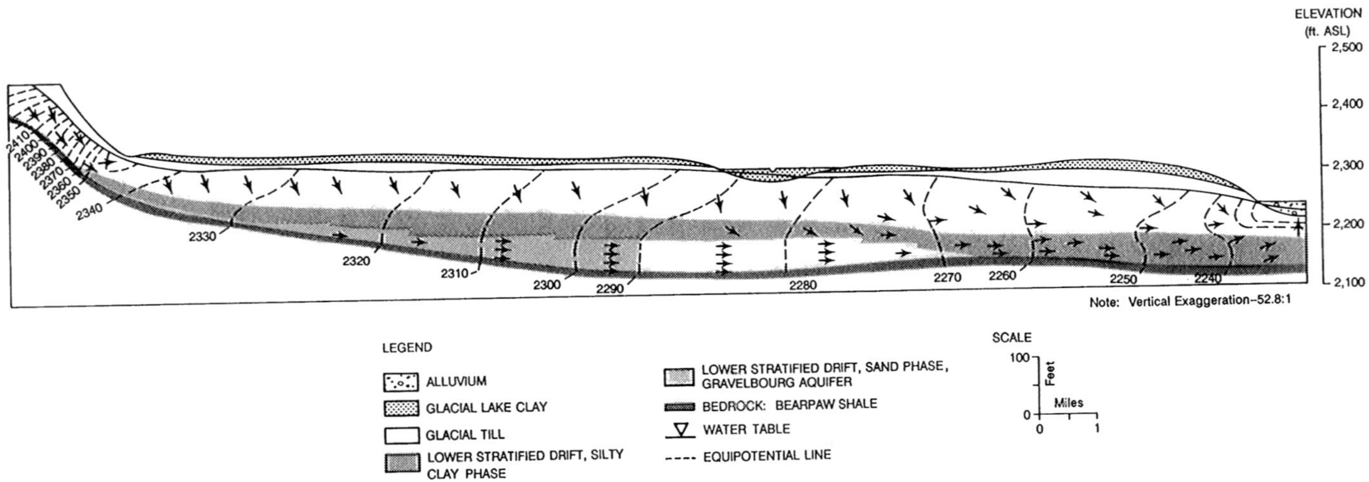

FIGURE 2.1 An example of a regional ground water flow system. The arrows depict the general direction of ground water flow in a potential field defined by the equipotential lines (modified from Freeze, 1969b).

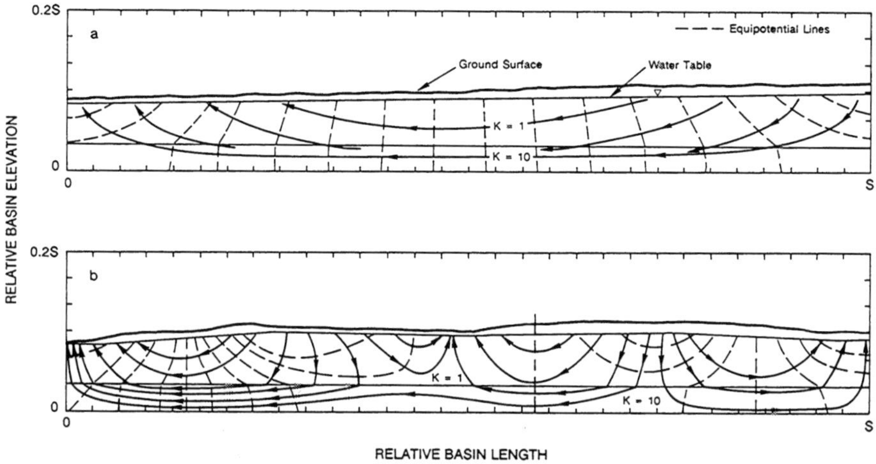

FIGURE 2.2 Dependence of the pattern of ground water flow on the recharge rate, as reflected by the configuration of the water table. All other parameters are the same in the two sections (from Freeze, 1969b).

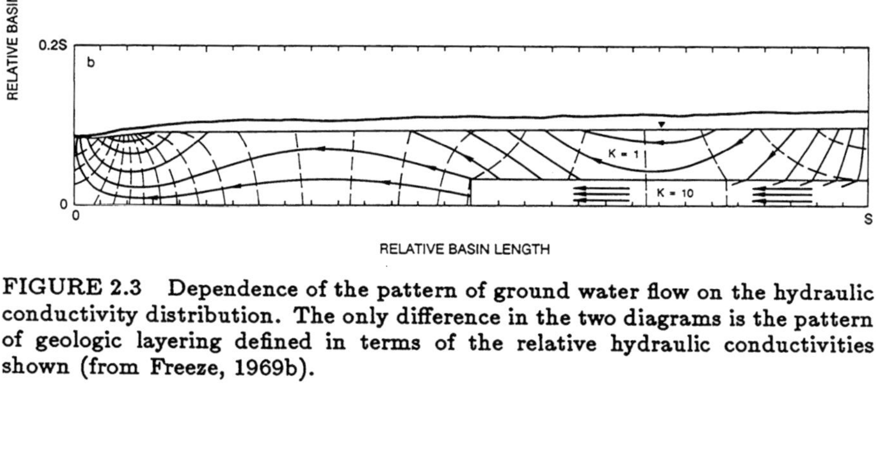

FIGURE 2.3 Dependence of the pattern of ground water flow on the hydraulic conductivity distribution. The only difference in the two diagrams is the pattern of geologic layering defined in terms of the relative hydraulic conductivities shown (from Freeze, 1969b).

hydraulic conductivity. Examination of the two flow patterns shows how changes in the hydraulic conductivity distribution can change the character of ground water flow.

Adding or removing water also can have a significant impact on the pattern of flow. The most important sources and sinks in a ground water flow system are pumping or injection wells (i.e., point sources/sinks). These are considered internal flows of water (fluxes). Other possibilities such as recharge or evaporation are most often considered as boundary fluxes. Pumping lowers the hydraulic potential at the well and in its immediate vicinity, creating what is known as a cone of depression. The result of decreasing hydraulic potential toward the well is the flow of water to the well. Injection does the opposite and results in flow away from a well.

So far, only steady-state flow, or flow that does not change as a function of time, has been discussed. Often, however, flow systems are transient, which means that hydraulic heads change with time, leading to variations in flow rates. For example, water levels decline when a pumping well is first turned on, providing an early transient response. In many instances when sources of recharge are available, water levels will eventually stabilize, providing a new equilibrium or steady-state flow system. The most important feature of a transient flow system is the ability of water to be removed from or added to storage in individual layers. The parameter describing the water storage capabilities of a geologic unit is called the "specific storage." For transient flow problems, its value contributes to determining the distribution of hydraulic head at a given time. Note that the smaller the specific storage, the faster the ground water system will seek a new equilibrium. Readers wishing a more detailed explanation of this parameter and aspects of ground water flow should see Freeze and Cherry (1979).

Multiphase Flow

Multiphase flow occurs when fluids other than water are moving in the subsurface. These other fluids can include gases found in the soil zone or certain organic solvents that do not appreciably dissolve in water (i.e., immiscible liquids). Examples of fluids that are immiscible with water include many different manufactured organic chemicals such as the cleaning solvent trichloroethylene and preservatives such as creosote. Petroleum products such as crude oil, heating oil, gasoline, or jet fuel are also examples.

The process causing all of these phases to flow is again movement in response to a potential gradient. Now, however, the situation is more complicated because the potential causing each fluid to move is not necessarily the same as that for water. Thus each fluid can be moving in a different direction and at a different rate. Another complexity is that many characteristic parameters are no longer constant when several fluids are present together and competing for the same pore space. For example, the relative permeability of a geologic unit to a particular fluid like water will be small if the proportion of water present in a given volume of porous medium is small and will tend to increase as the amount of water increases.

As discussed previously for water, a fluid's potential also depends on any sources or sinks that add or remove fluid. The same idea applies to multifluid systems, except that now the number of processes increases because the effects of pumping/injection and evaporation (volatilization) affect each of the fluids present and, in addition, there can be transfers of mass between fluids. An example of this latter mechanism is that some portion of a gasoline spill might dissolve in water.

To illustrate these concepts about the theory of multiphase systems, consider two problems of particular interest to this report—the flow of water in the unsaturated zone and the migration of organic contaminants that are either more or less dense than water. When studying the problem of water movement in the presence of soil gas in the unsaturated zone, it is sometimes assumed that only the water moves. The only effect of the gas on water movement is the variability in the parameters caused by the presence of several fluids. For example, hydraulic conductivity varies as a function of the quantity of water in the pores.

Figure 2.4 shows a relationship between hydraulic conductivity (K) and pressure head (ψ). According to Freeze and Cherry (1979), pressure head is one component of the total energy water possesses at a point. Several features should be noted. As the pressure head becomes smaller (more negative), the soil becomes drier and the hydraulic conductivity decreases. Much less water will move through a dry soil than through a wet soil. Another feature is that if the soil is drying out there is one ψ-K relationship and if it is wetting there is another. Further, repeated wetting and drying cause the relationship to be defined by the scanning curves that join the wetting and drying curves at intermediate points (Figure 2.4). In

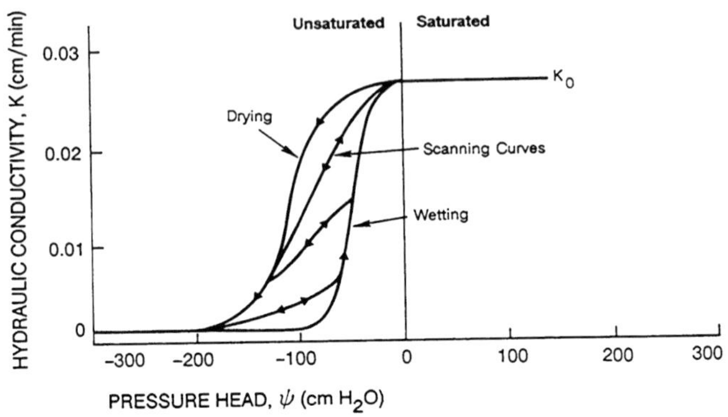

FIGURE 2.4 Example of the relationships between pressure head and hydraulic conductivity for an unsaturated soil (modified from Freeze, 1971a).

most multifluid systems, hydraulic conductivity and other parameters commonly exhibit this kind of "hysteretic" behavior, and yet for many applications, these types of site-specific data are not available.

The progress of a wetting front moving into a dry soil can be described in terms of either potentials or volumetric water content,

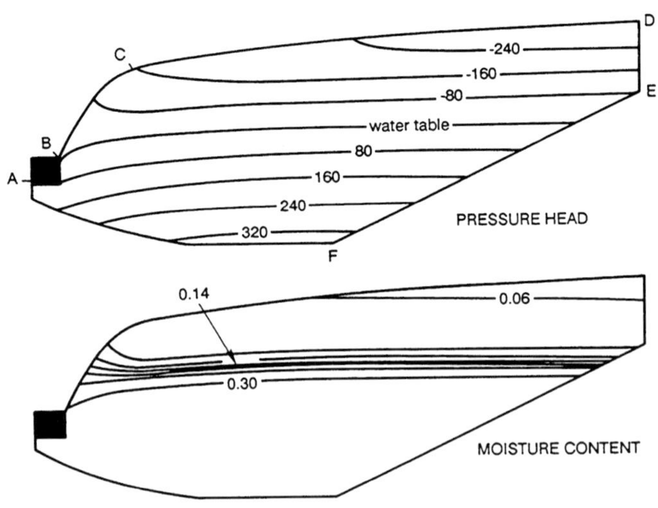

FIGURE 2.5 The distribution of water in the unsaturated zone can be described in terms of pressure head and moisture content. Results presented for a combined saturated and unsaturated flow system illustrate how pressure head in particular is continuous across the water table (from Freeze, 1971b).

defined as the ratio of the volume of water in the voids to the total volume of voids. Figure 2.5 illustrates how both are used to define a wetting zone near the top of the ground. The water table is clearly illustrated by the zero pressure contour and the total porosity contour (complete saturation). Given that moisture contents are easier to measure than potentials, the former are used more frequently to describe real systems.

A more complicated case to consider is a flow involving an immiscible fluid and water in the subsurface. Eventually, a distinction has to be made between a fluid that is less dense than water and one that is more dense. However, where an organic liquid is spilled on the ground surface, both fluids will move much the same way through the unsaturated zone (Figure 2.6a and b). The free organic liquid in a homogeneous medium moves vertically downward, leaving a residual trail of organic contaminants. Each pore through which the free organic liquid moves retains some of the contaminant (residual saturation) in a relatively immobile state. Thus, if the volume of spilled liquid is small and the unsaturated zone is relatively thick, no free liquid may reach the water table. Of course, free liquid may reach the water table over extended periods of time, and dissolved organic liquid may be conducted by water flow.

It is when the free liquids begin to approach the top of the capillary fringe above the water table that the differences in density begin to affect transport. The capillary fringe is a zone above the water table where the pores are completely saturated with water but the pressure heads are less than atmospheric (Freeze and Cherry, 1979). A contaminant that is lighter than water will mound and spread, following the dip of the water table (Figure 2.6a). A fluid that is heavier than water will spread slightly and keep moving downward. This fluid will ultimately mound on the bottom of the aquifer or on a low-permeability bed within the aquifer and move in whatever direction the unit is dipping (Figure 2.6b). Thus water and the organic liquid need not move in the same direction.

To understand the details of multicomponent flow, it is essential to study the concepts of wettability, imbibition and drainage, and relative permeability. A discussion of these topics is, however, beyond the scope of this report. Readers can refer to Bear (1972) and Greenkorn (1983) for an overview of the basic theory. The key point to remember is that, as in the case of water, the permeability of the material through which these fluids are moving plays a major role in controlling the direction and rate of flow. In the case of multiphase

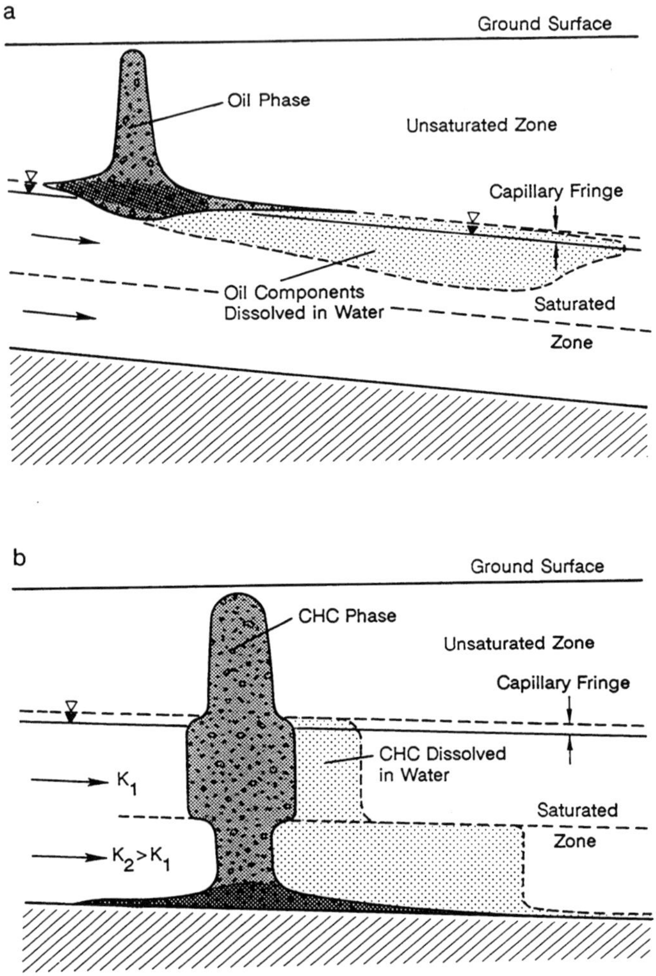

FIGURE 2.6 The flow of a nonaqueous-phase liquid that is (a) less dense than water (oil) and (b) more dense than water (chlorohydrocarbon, CHC) in the unsaturated and saturated zones. In both cases the contaminants are also transported as dissolved compounds in the ground water (from Schwille, 1984).

systems, a relative permeability is defined for each fluid with values ranging between zero and one as the relative abundance of each fluid (i.e., saturation) changes. The key point here is that as more fluids are introduced to the pore space, more of the pore space is devoted to the relatively immobile state of each fluid and therefore less pore space is devoted to liquid flow.

The distribution of an NAPL in the subsurface is described quantitatively in terms of the relative saturation of the NAPL, which is given by the ratio of the volume of the NAPL to the total pore volume. In other words, it describes what proportion of the pore volume is filled with the NAPL. A relative saturation can be defined for each one of the organic liquids and water. This kind of description is generally not used in field settings because of the detailed study that is necessary. Instead, presence/absence indications are used, as illustrated in Figure 2.6. The results of computer simulations normally characterize relative fluid saturations. Expressing the distribution of fluids in terms of relative saturation is analogous to expressing the moisture content in terms of unsaturated flow.

Dissolved Contaminant Transport

One of the reasons why problems involving dissolved contaminants are so difficult to model is the number and complexity of controlling processes. The processes can be divided into two groups: (1) those responsible for material fluxes and (2) sources or sinks for the material. For the problem of contaminant migration these are the mass transport and mass transfer processes, respectively (Table 2.1). A brief discussion of each of the processes listed in Table 2.1 follows, with a general assessment of its impact on contaminant transport.

Advection

Advection is the primary process responsible for contaminant migration in the subsurface. Mass is transported simply because the ground water in which it is dissolved is moving in a flow system. In most cases, it can be assumed that dissolved mass is transported in the same direction and with the same velocity as the ground water itself. For example, given the conditions of flow described by the equipotential lines and flowlines of Figure 2.7a, it is a simple matter to define the plume of dissolved contaminants in terms of the streamtubes that pass through the source. A streamtube is defined as the area between two adjacent flowlines. When flowlines are equally spaced, the discharge of water through each is the same (Freeze and Cherry, 1979). This simple approach assumes that the density of the contaminated fluid is about the same as that of the ground water. The mean velocity of contaminant migration can also be assumed to be the same as the mean ground water velocity (or seepage velocity).

TABLE 2.1 A Summary of the Processes Important in Dissolved Contaminant Transport and Their Impact on Contaminant Spreading

Process	Definition	Impact on Transport
Mass transport		
1. Advection	Movement of mass as a consequence of ground water flow.	Most important way of transporting mass away from source.
2. Diffusion	Mass spreading due to molecular diffusion in response to concentration gradients.	An attenuation mechanism of second order in most flow systems where advection and dispersion dominate.
3. Dispersion	Fluid mixing due to effects of unresolved hetero-geneities in the per-meability distribution.	An attenuation mechanism that reduces contaminant concentration in the plume. However, it spreads to a greater extent than predicted by advection alone.
Chemical mass transfer		
4. Radioactive decay	Irreversible decline in the activity of a radionuclide through a nuclear reaction.	An important mechanism for contaminant attenuation when the half-life for decay is comparable to or less than the residence time of the flow system. Also adds complexity in production of daughter products.
5. Sorption	Partitioning of a contaminant between the ground water and mineral or organic solids in the aquifer.	An important mechanism that reduces the rate at which the contaminants are apparently moving. Makes it more difficult to remove contamination at a site.
6. Dissolution/ precipitation	The process of adding contaminants to, or removing them from, solution by reactions dissolving or creating various solids.	Contaminant precipitation is an important attenuation mechanism that can control the concentration of contaminant in solution. Solution concentration is mainly controlled either at the source or at a reaction front.
7. Acid/base reactions	Reactions involving a transfer of protons (H^+).	Mainly an indirect control on contaminant transport by controlling the pH of ground water.

TABLE 2.1 *Continued*

Process	Definition	Impact on Transport
8. Complexation	Combination of cations and anions to form a more complex ion.	An important mechanism resulting in increased solubility of metals in ground water, if adsorption is not enhanced. Major ion complexation will increase the quantity of a solid dissolved in solution.
9. Hydrolysis/ substitution	Reaction of a halogenated organic compound with water or a component ion of water (hydrolysis) or with another anion (substitution).	Often hydrolysis/substitution reactions make an organic compound more susceptible to biodegradation and more soluble.
10. Redox reactions (biodegradation)	Reactions that involve a transfer of electrons and include elements with more than one oxidation state.	An extremely important family of reactions in retarding contaminant spread through the precipitation of metals.
Biologically mediated mass transfer		
11. Biological transformations	Reactions involving the degradation of organic compounds, whose rate is controlled by the abundance of the microorganisms and redox conditions.	Important mechanism for contaminant reduction, but can lead to undesirable daughter products.

The close relationship between advective transport and ground water flow means that the factors considered for flow, the location and quantity of the inflow and outflow to the flow system, the hydraulic conductivity distribution, and the presence of pumping/injection wells also play a major role in determining where contaminants migrate. Indeed, the process of advection is often so dominant that the mean velocity predicted by flow models can be used to estimate patterns of contaminant transport with surprising accuracy.

Diffusion

Diffusion is an important process that results in mass mixing. Diffusion is mass transport in response to a concentration gradient. Thus contaminants present in a plume will diffuse away from

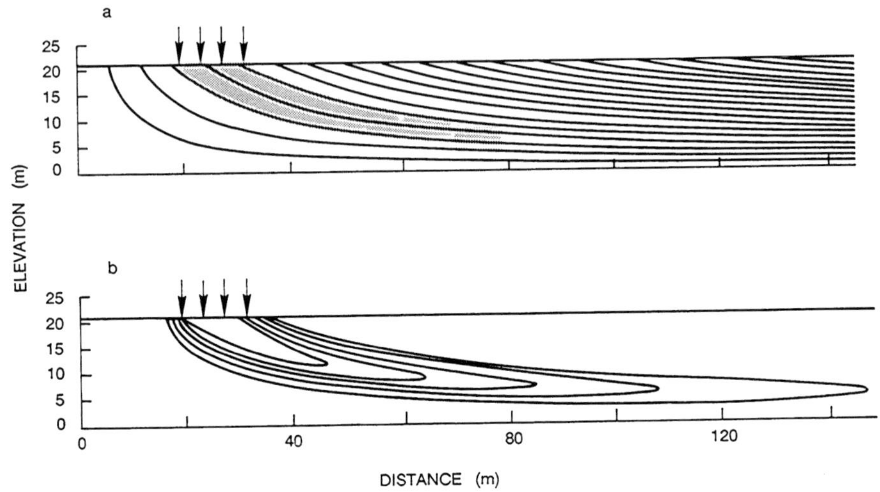

FIGURE 2.7 Plume produced (a) by advection alone and (b) by advection and dispersion (adapted from Frind, 1987).

the plume in all directions in response to concentration gradients. Although this process occurs in most contaminant problems, its overall contribution to the spreading of contaminants is usually negligible. There are situations, mainly in fractured rock settings and low-permeability units, where diffusive mass transport is of primary importance and needs to be considered.

Dispersion

Dispersion refers generally to phenomena that cause fluid mixing. Dispersion is more accurately described as the apparent mixing due to unresolved advective movement at scales finer than captured by the mean advection model. Essentially, dispersion produces a mixing zone between the contaminated water and the native ground water. This effect can be illustrated by considering a plume developed due to advection alone (Figure 2.9a) and modifying it to also include dispersion (Figure 2.7b). A comparison of parts a and b in Figure 2.7 shows that dispersion has expanded the plume size beyond that expected due to advection alone. Contaminants spread into adjacent streamtubes and farther down the streamtube where the contaminants are migrating. The overall plume becomes larger, and, in general, the concentration is less than was the case with advection alone.

Dispersion in a direction perpendicular to the mean direction of ground water flow is termed transverse dispersion, while dispersion parallel to the mean direction of flow is termed longitudinal dispersion. There are actually two directions of transverse spreading (Figure 2.8). These different components of dispersion usually need to be considered separately in models because spreading upward or downward is often considerably less than spreading in horizontal or subhorizontal planes.

Hydrodynamic mixing occurs as a consequence of nonidealities at various scales that result in local variability in velocity around some mean velocity. For example, at the scale of pores (Figure 2.9a) this variability may be caused by velocity variations within a pore or by subtle changes in the flow network that cause the mass to spread out or finger into adjacent pores of the pore network. At the macroscopic scale, the variability can be due to heterogeneities in the hydraulic conductivity distribution (Smith and Schwartz, 1980) of the kind shown in Figure 2.9. In terms of the relative magnitude or mixing due to hydrodynamic dispersion and diffusion, the former is by far the more significant. Readers interested in learning more about dispersion should refer to Schwartz (1975, 1977), Anderson (1979, 1984), Tennessee Valley Authority (1985), Mackay et al. (1986), Freyberg (1986), and Sudicky (1986).

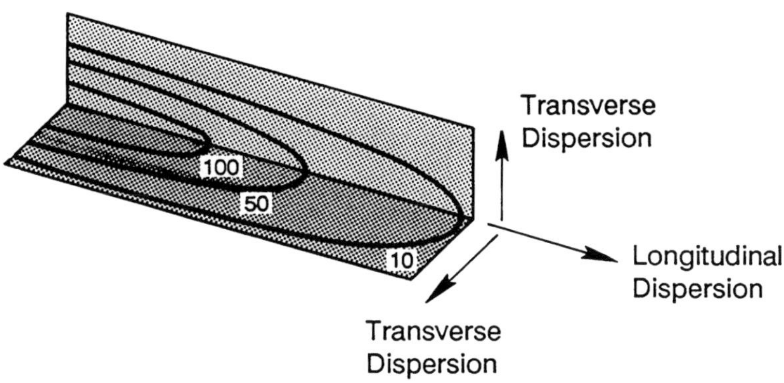

FIGURE 2.8 Idealized pattern of plume spreading in three dimensions is characterized by a longitudinal and two transverse dispersion components.

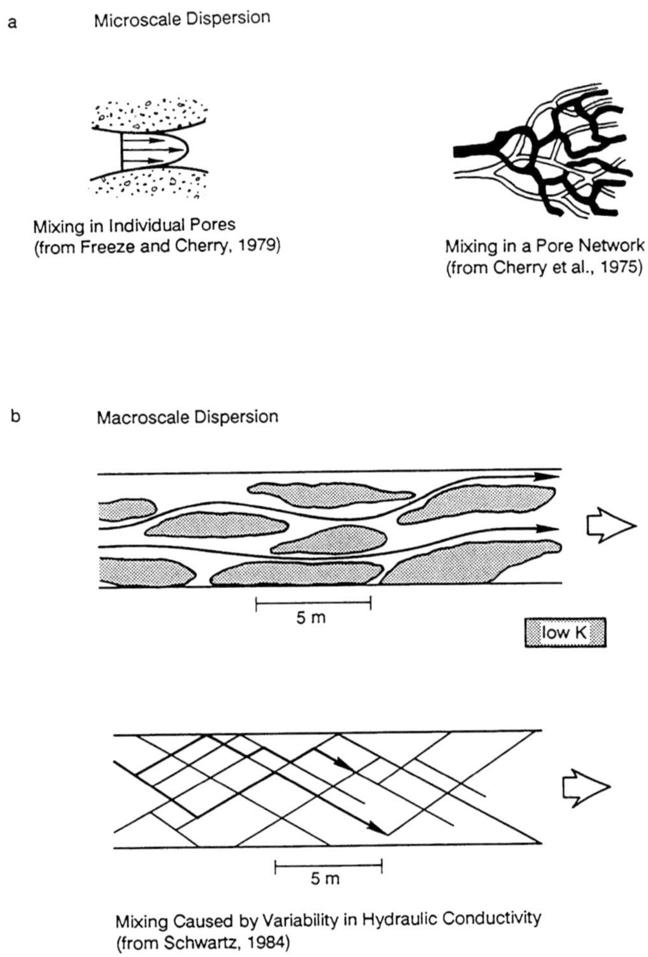

FIGURE 2.9 (a) Microscale and (b) macroscale variability contributing to the development of dispersion.

Radioactive Decay

Radioactive decay, the transformation of one element into another through the loss of atomic particles from the element's nucleus, is a process that has been thoroughly characterized and is well understood. Radioactive decay leads to the loss of the original radioactive isotope from ground water over a period of time, but daughter products are produced that may also be of environmental concern. A simple rate law can be applied to the decay of any radioactive isotope, and it has been included in various transport models for many years. It describes an exponential decrease with time in the concentration of the dissolved radioactive component.

The best-known example of radioactive decay in ground water is that of tritium (^3H). Tritium is produced naturally by interaction of cosmic rays (various nuclear particles coming in from outer space) with gases in the upper atmosphere. Consequently, trace amounts of tritium are found in all natural waters. During the 1950s, large amounts of new tritium were injected into the atmosphere as a result of the testing of fusion bombs. The anomalously high concentrations of bomb-produced tritium led to much interesting work in dating and tracing the patterns of flow of ground waters. Tritium, with a half-life of 12.5 yr, decays to stable helium (^3He) by emission of a beta particle. Because of its short half-life, tritium produced by the testing of atomic weapons in the atmosphere is gradually disappearing from natural waters.

Of environmental importance are radioactive species that may be inadvertently released into ground water from such activities as mining, milling, and storage of wastes. In particular, concern exists about the escape and potential hazards of radium, uranium, and lead in ground water adjacent to uranium mills and processing plants, and about the leaching and movement of radioactive isotopes (including isotopes of uranium, plutonium, cesium, neptunium, europium, iodine, selenium, and others) away from geologic repositories for high-level radioactive wastes from commercial power plants and defense installations (Bates and Seefeldt, 1987; Fried, 1975). Because the radioactive isotopes of concern in the environment undergo other chemical reactions in addition to radioactive decay, many years of research will be required before their behavior can be modeled with confidence, even though radioactive decay is well understood.

When the half-life for radioactive decay is of the same magnitude or smaller than the residence time of the contaminant in the subsurface, decay significantly affects contaminant migration. This is illustrated in Figure 2.10. Figure 2.10a illustrates what a hypothetical plume might look like if advection and dispersion were the only controlling processes. The plume is much larger than the one in Figure 2.10b, where it is assumed that radioactive decay is also operative.

As chemical and biological processes are discussed, it will become apparent that in general their effect is to attenuate the spread of

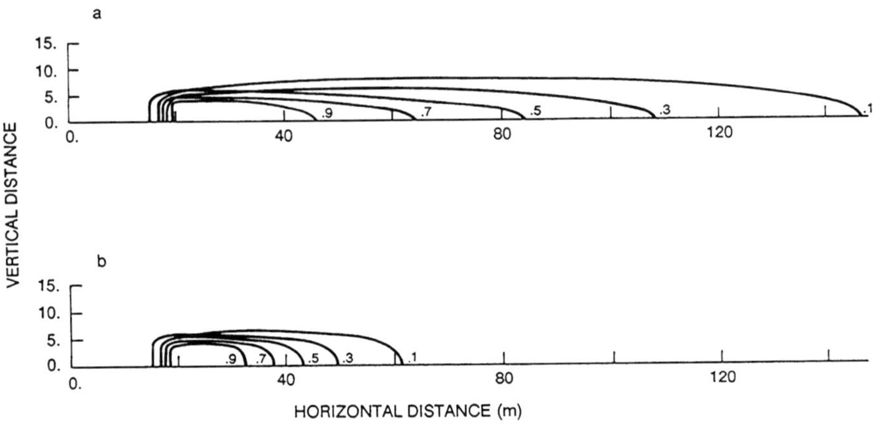

FIGURE 2.10 Many of the geochemical processes like radioactive decay and sorption attenuate the spread of contaminants. Compared on the figure are map views of half-plumes (a) without attenuation and (b) with attenuation (based on Frind, 1987).

contaminants relative to that caused by advection and dispersion alone. It is for this reason that so much emphasis in modeling has been placed on accounting for these processes to the fullest extent possible.

Sorption

Adsorption reactions remove contaminants from ground water and add them to the surfaces of minerals or the solid organic carbon of the unit through which the contaminants are moving. The term sorption is a general one that includes adsorption (attraction to a surface), absorption (incorporation into the interior of a solid), ion exchange (adsorption, with a charge-for-charge replacement of the ionic species on a surface by other ionic species in solution), and desorption (the opposite of each of the above adsorption reactions). Sorption will affect virtually all dissolved species in ground water to some degree.

Sorption is such a complex process that it is not really possible within the scope of this review to provide a complete appreciation of what causes contaminants to move from solution onto solids. Metal ions are sorbed primarily because of the positive charges they carry or chemical reactions that bind them to the surface. Clay minerals in particular have large surface areas carrying an overall negative charge. This surface charge is balanced by positively charged ions that are attracted to the surface. Contaminant ions in ground water can in many cases preferentially replace these positively charged species. Other surfaces (e.g., metal oxides or metal oxyhydroxides) are reactive in the sense that a metal or certain metal-containing compounds can be chemically bound to the surface. These surfaces are particularly interesting because the reactivity of the surface, or the ability to sorb contaminants, is controlled in part by the pH of the ground water.

Another major class of contaminants, noncharged organic molecules, sorb mainly onto solid organic material. The force that drives the exchange in this case is the hydrophobic (water hating) character of some organic compounds. For such chemicals, sorption onto solids increases as the solubility in water decreases. Thus the more water soluble the contaminant, the less likely it is to be sorbed. When an organic phase is present, either solid or liquid, organic contaminants prefer to reside in that phase.

Many sorption reactions discussed are completely or partially reversible. In other words, if the concentration of the contaminant in the ground water decreases, desorption will occur to maintain an equilibrium between the contaminant in solution and that sorbed on the solids. Thus sorption does not permanently remove a contaminant from solution but instead only stores it.

A number of empirical or semiempirical methods have been developed for describing sorption equilibrium. The so-called distribution coefficient (K_d) is the most simple of these; K_d is defined as the concentration of a given contaminant sorbed on the solid phase (commonly in micrograms per gram) divided by the concentration of the same contaminant in solution (in micrograms per milliliter), with the resulting units being milliliters per gram. A large K_d value indicates strong sorption, or that the compound distributes itself primarily onto or into the solid phase. A small K_d indicates that the compound stays mainly in the water phase. Therefore K_d serves as a qualitative guide to the relative tendency toward sorption of various dissolved species in a given solution. The usefulness of K_d is diminished somewhat by the fact that it may vary as a function of concentration, ionic strength, competing ions, and other factors.

In soils, sediments, and some aquifers, solid organic matter is the primary solid material onto which organic compounds sorb. For many organic compounds, empirical relationships can be derived to predict the K_d value as a function of the amount of organic material in the solid phase and a measure of the organic compound's hydrophobic nature, most usually its octanol/water partition coefficient. A large octanol/water partition coefficient signifies a highly hydrophobic compound, which will have a large K_d.

The overall effect of sorption is to retard or delay the spread of contaminants. This effect is not unlike that shown in Figure 2.10. When sorption occurs, the rate at which the contaminant appears to move is lower than would be the case for an unretarded or neutral tracer. This behavior helps to reduce the spread of contaminants but also makes it more difficult to remove contamination from the ground; that is, it tends to increase the time required to remediate to a cleanup level.

Precipitation and Dissolution

Dissolved contaminants can be either lost from solution or brought into solution by the processes of precipitation and dissolution of a solid phase. Runnells (1976) gives examples of precipitation of dissolved contaminants in ground water caused by reactions with other dissolved species, hydrolysis, and reduction or oxidation. Examples of contaminants that could be reduced to lower concentrations in ground water through the formation of precipitates include arsenic (by reaction with iron, aluminum, or calcium), lead (by reaction with sulfide or carbonate), and silver (by reaction with sulfide or chloride). Hydrolysis can lead to the precipitation of iron, manganese, copper, chromium, and zinc contaminants. Oxidation or reduction could favor the precipitation of chromium, arsenic, and selenium.

The precipitation of dissolved contamination plays an important role in contaminant attenuation. Although this process is not well described in case studies, theoretical work shows that it will attenuate the spread of contaminants by removing mass from solution as saturation is exceeded. Unlike the sorption processes that also partition mass between the solid and solution, these reactions are less reversible. For example, metals precipitating as metal-sulfides are virtually immobilized for as long a time as the general chemical environment remains constant.

Contaminant dissolution is an important reaction that can occur at a source to initially bring contaminants into solution. However, for some minerals that dissolve relatively rapidly, it is possible for contaminants initially immobilized by precipitation to be remobilized as the plume moves further down the flow system. A natural analog of the repeated precipitation and remobilization of metals is a uranium roll-front deposit (Galloway and Hobday, 1983).

Acid/Base Reactions

Reactions involving the gain or loss of the hydrogen ion (H^+) are called acid/base reactions. Acids are chemical species that give up or donate a H^+ ion, while bases are species that accept a H^+ ion. Many potential contaminants are susceptible to change in speciation because of changes in pH. For example, under oxidizing conditions, dissolved arsenic should be present in normal ground water (with a pH of 7 to 8) in the form of $HAsO_4{}^{2-}$. However, if the pH of the water is lower than about 6, the dominant form of oxidized arsenic is $H_2AsO_4{}^-$ or, at very low pH, $H_3AsO_4{}^-$. Depending on the number of protons attached to the arsenate, the chemical behavior of the arsenic in solution may be quite different. For example, the sorption behavior of H_3AsO_4 is quite different from the sorption behavior of $H_2AsO_4{}^-$.

There are cases where the chemical reaction being considered does not include the contaminant. In terms of understanding transport, this means that in some cases the reactions that control important geochemical parameters of a system (e.g., pH) must be considered even though the compounds or ions involved in the specific reaction are not contaminants. For this reason, sophisticated mass transport models often need to include reactions related to the CO_2-water system, one of the dominant controls of the pH of ground water.

Complexation

The process of complexation is the combination of simple cations and anions into more complex aqueous species. According to Morel (1983), complexes can be classified as ion pairs of major constituents, inorganic complexes of rare metals, and organic complexes. Following are examples of reactions forming each of these complexes:

$$Ca^{2+} + SO_4^{2-} = CaSO_4,$$

$$Cu^{2+} + H_2O = CuOH^+ + H^+$$

$$Cu^{2+} + Y^- = CuY^+,$$

where Y^- is an organic species such as glycine. Conventionally, complexation reactions are modeled as equilibrium processes where equations like those above are characterized in terms of mass law expressions and equilibrium constants. This relatively straightforward treatment enables the concentration of the individual complexes in water to be easily calculated.

In terms of mass transport, complexation reactions are important mainly because of the role they play in increasing the mobility of metals. For example, over the range of pH common to most natural ground waters, metals (present as ions) will occur at relatively low concentration because of the solubility constraints provided by solid phases including metal-hydroxides, carbonates, or sulfides. However, when metals complex to a significant extent, the total quantity of a particular metal dissolved in water can be much larger than simply the concentration of the metal ion itself. Overall then, complexation can enhance the quantity of metals being moved in a contaminant plume. It is for this reason that most models involving metal transport need to consider complexation reactions.

Another instance where complexation needs to be considered is in the sorption of metals on surfaces whose charge changes as a function of pH. Examples of such surfaces include kaolinite, metal oxides, and metal oxyhydroxides. For such solids, the sorptive behavior changes depending upon which metal species (ion and complexes) are present in the solution. Additional complexity arises from the fact that changing the composition of the water (e.g., pH) also changes the concentration of various metal species in solution. Thus, in situations where this type of sorption can occur, the metal complexation must be included to fully characterize the sorption reactions.

Hydrolysis/Substitution

Hydrolysis and substitution are abiotic transformation reactions that affect organic contaminants in ground water. The term hydrolysis refers specifically to substitution reactions involving water or a component of water, for example:

$$RX + H_2O \rightarrow ROH + HX,$$

where R refers to the main part of the organic molecule and X is a halogen (e.g., Cl^-, Br^-) (Jackson et al., 1985). However, not all substitution reactions involve water. For example HS^- can react and substitute for a halogen (e.g., Br^-) in a reaction of the following kind (Jackson et al., 1985):

$$RCHX + HS^- \rightarrow RCHSH + X^-.$$

Kinetic rate laws for hydrolysis/substitution reactions can be complex. However, in some cases, they can be approximated as first-order reactions or, in other words, reactions like radioactive decay that can be described simply in terms of a half-life. The reason transformation reactions are important is that products are often more susceptible to biodegradation and more soluble.

Redox Reactions

Any element that can have different valences is potentially subject to transformation via oxidation and reduction reactions, known in short as "redox" reactions. Included are abundant species, such as nitrogen, sulfur, carbon, and phosphorus, and minor trace species, such as iron, manganese, uranium, selenium, copper, and arsenic. Redox reactions involve the movement of electrons from one species to another. Redox reactions change the speciation of the dissolved elements, and they can result in the removal of a dissolved element when the product species is involved in a phase transfer reaction. As an example of a redox reaction, consider the oxidation of Fe^{2+}, described by the following reaction:

$$O_2 + 4Fe^{2+} + 4H^+ = 2H_2O + 4Fe^{3+}.$$

In this reaction, the exchange of electrons changes the oxidation number of oxygen from (0) to (−II) and iron from (+II) to (+III). For such reactions, there are reductants (electron donors, e.g., Fe^{2+}) and oxidants (electron acceptors, e.g., O_2).

Fundamental problems exist in the conceptualization of redox reactions. Most explanations make the assumption that the reactions among aqueous species are reversible and at equilibrium. However, it is abundantly clear from various lines of evidence (see, for example, Lindberg and Runnells, 1984) that many redox reactions are essentially irreversible. That is, the reactions may go in one direction easily but cannot be reversed (without biological intervention) to go in the other direction. Well-known examples include the reaction at low temperatures between such reduced-sulfur species as HS^- or S^{2-} and oxidized sulfur in the form of $SO_4{}^{2-}$; the oxidation to SO_4^{2-} readily occurs in the presence of oxygen, but the reduction of SO_4^{2-} cannot take place unless sulfate-reducing bacteria are actively involved. Similar examples of irreversible reactions are known for selenium, arsenic, nitrogen, and many other elements. Much research is needed to identify which redox reactions can be modeled as reversible reactions and which are irreversible and should not be included in an equilibrium model.

Biological Transformations

When organic and some inorganic compounds are present as contaminants, biological transformation can be an important process, because the original contaminant is destroyed. The advantages of biological transformation are two: (1) the contaminant can be completely mineralized to innocuous products, and (2) the process is not saturated, as with sorption, exchange, or filtration. Microorganisms can be involved with redox reactions, as described in the previous section, or with substitution and hydrolysis reactions. The microorganisms produce enzymes that allow the reactions to proceed much more rapidly. Therefore the kinetics of microbially mediated reactions are faster than those of the same reactions in the absence of the microorganisms.

The metabolic capabilities of the microorganisms that can exist in soils and aquifers are quite diverse and can allow biodegradation of almost all types of organic contaminants. For instance, the fungi are known for their ability to degrade complex polysaccharides and other polymers of natural origin. Their capability to degrade xenobiotic (i.e., man-made) chemicals is, however, thought to be small.

Bacteria have wide-ranging capability to degrade natural and xenobiotic organic compounds. Significant advances have been made over the past 10 years in elucidating the broad capabilities of bacteria toward xenobiotic chemicals. Table 2.2 lists classes of xenobiotic organic compounds that are known to be degraded by aerobic bacteria, while Table 2.3 lists compounds that are degraded by strictly anaerobic bacteria. For further information, several more thorough reviews can be consulted (Alexander, 1985; Atlas, 1981; Rittmann et al., 1988). Current research is being directed toward further defining the capabilities of bacteria for degradation of xenobiotics.

Biological reactions are driven by the ultimate goal of producing new cell mass. In order to accomplish this goal, microorganisms must transform environmentally available nutrients to forms that

TABLE 2.2 Classes of Xenobiotic Organic Compounds Known to Be Biodegraded by Aerobic Bacteria

Unsubstituted aromatics (phenols, benzenes, benzoates)
Nitro-substituted aromatics
Halogen-substituted aromatics
Polycyclic aromatics
PCBs
Most pesticides
Phthalate esters

TABLE 2.3 Classes of Xenobiotic Organic Compounds Known to Be Biodegraded Under Strictly Anaerobic Conditions

Halogenated aliphatic solvents
Most unsubstituted aromatics
Some PCBs
Some pesticides

are useful for incorporation into cells. Then they must synthesize the useful components into the polymers that make up the cell mass.

The environmentally available nutrients often are not in the form needed by the cells. In general, cells utilize reduced forms: e.g., NH_4^+–N, HS^-–S, and CH_2O–C. However, the commonly available forms often are oxidized, e.g., NO_3^-–N, SO_4^{2-}–S, and CO_2–C. In order that the needed forms can be made, a source of electrons is necessary. Hence an essential feature for growing cells is having an electron donor.

Reducing nutrients takes energy as well as electrons. Synthesizing the polymers that make up cells, repairing or replacing cell constituents, transporting nutrients across the cell membrane, and motility are also significant energy sinks. Therefore cells must have an energy source if they are to grow and sustain themselves. In most cases the energy source is the electron donor.

To generate energy the electron donor must donate its electrons to an electron acceptor, making available the energy for cell synthesis.

Following is an example of a biologically mediated redox reaction in which an organic compound typified as CH_2O is oxidized to simpler compounds:

$$(1/4)CH_2O + (1/4)O_2 = (1/4)CO_2 + (1/4)H_2O.$$

In this reaction, the transfer of electrons between the organic compound CH_2O (the electron donor) and O_2 (the electron acceptor) provides the energy required for cell growth. In addition to oxygen, other potential electron acceptors could include NO_3^-, NO_2^-, SO_4^{2-}, CO_2, and certain organic compounds.

To summarize, growth of cells requires environmentally available nutrients, an electron donor, an electron acceptor, and an energy source. If all these factors are present and if the environment is not toxic, microbial growth is possible. Whether or not a contaminant of interest fulfills any of these needs, some materials must fill the need if microorganisms are to accumulate in the environment.

WHAT IS A MODEL?

A mathematical model is a replica of some real-world object or system. It is an attempt to take our understanding of the process (conceptual model) and translate it into mathematical terms. Therefore the mathematical model is only as good as our conceptual understanding of the process. A mathematical model differs from other models (e.g., physical, analog) in its attempt to simulate the actual behavior of a system through the solution of mathematical equations. In this sense, a mathematical model is much more abstract than physical models.

The three main components of a model are the specific information describing the system of interest (e.g., what processes are important), the equations that are solved in the model, and the model output. A requirement for solving the equations embedded in a model is data about the user's particular problem. These data include specified numerical values for parameters describing the processes, for simulation parameters that are part of the procedure used to solve the equations, and for parameters describing the shape or geometry of the region of interest. This information provided by the user customizes the model to the particular problem. These input data in conjunction with the governing equations determine system behavior under the specified conditions.

As an example of the steps in modeling, consider a ground water flow problem (Figure 2.11). Information that needs to be provided to describe a real system (Figure 2.12a) could include the following:

- the shape of the modeled region,
- the hydraulic conductivity and specific storage distributions in that region,

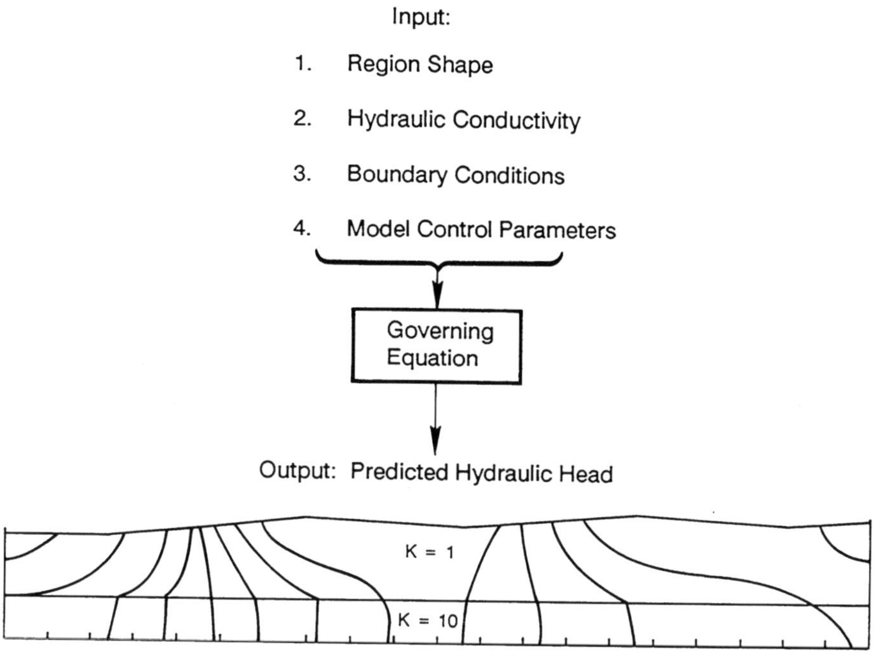

FIGURE 2.11 The components of a model: input data, a governing equation solved in the code, and the predicted distribution, which for this example is hydraulic head.

- the boundary and initial conditions, and
- model control parameters.

The model control parameters represent information like grid size and time step sizes that is necessary for the numerical solution of the differential equations. These details are used, given the capability of the model, to solve the partial differential equation(s) describing ground water flow. Output from the model is a predicted hydraulic head distribution for the specified region (Figure 2.12b).

Each of the different types of flow problems of interest (e.g., ground water flow, multiphase flow, and contaminant transport) will have a different governing equation, reflecting the fact that different processes are involved in ground water flow in comparison to other processes such as dissolved contaminant transport. Not unexpectedly, then, the information that a user supplies will also be different because the various processes have unique parameters. The model control parameters will also vary from one model to another because,

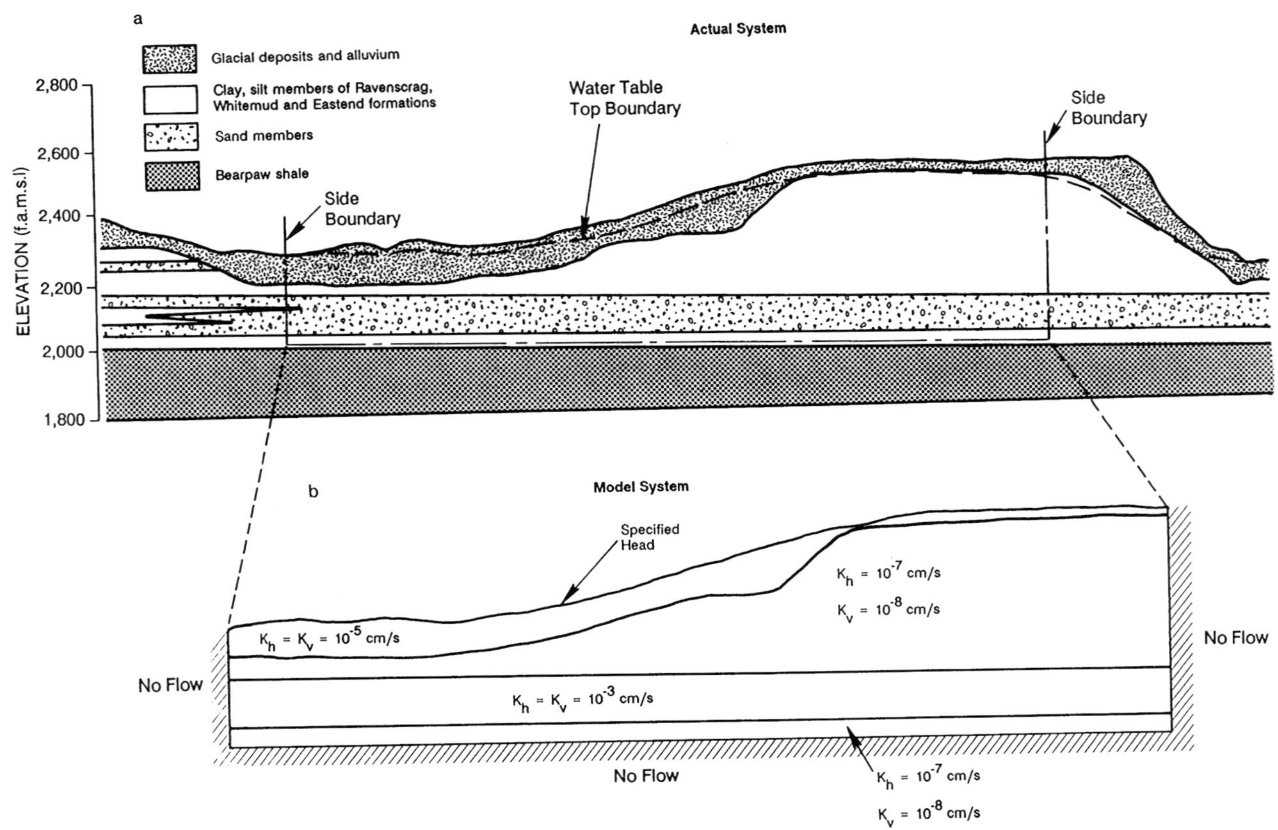

FIGURE 2.12 An example of how (a) a "real" system is represented by (b) a model system, which is defined by a region shape, boundary conditions, and hydraulic parameters. The example section comes from Freeze (1969a).

in general, many different mathematical techniques are available to solve a given equation. The following sections examine these issues, beginning with the governing equations and how the processes are actually incorporated. Next, some of the different methods available to solve flow and transport equations are discussed.

Governing Equations

The development of the flow and contaminant transport equations is relatively straightforward because all of the flow problems of interest here—ground water flow, multicomponent flow, and dissolved contaminant transport—are developed from the same fundamental principle, namely, the conservation of fluid or dissolved mass. Given a block of porous medium, the general conservation equation for the volume can be expressed as

$$\text{rate of mass input} - \text{rate of mass output} + \text{rate of mass production/consumption} = \text{rate of mass accumulation.} \quad (2.1)$$

The differential equations for flow or mass transport are simply mathematical expressions of this conservation statement incorporating the relevant processes. For fluid flow, the process responsible for moving fluid mass into and out of the volume element is simply flow in response to a potential gradient, while the rates of mass production or consumption include processes such as injection or pumping that add or remove fluid mass directly to or from the volume element. The same ideas hold for mass transport, with advection and dispersion responsible for moving mass into or out of the volume element, and the chemical and biological processes acting as source-sink terms.

The details of how the governing equations are developed are relatively complicated and are best left to textbooks (Bear, 1972, 1979; Freeze and Cherry, 1979; Greenkorn, 1983). A brief overview of how the governing equations are formulated is included here. However, those not interested in the mathematics or the details of how the equations are developed should simply be aware that equations can be developed that describe flow problems of interest; they can continue the general overview of models with the section "Solving Flow and Transport Equations."

As a simple example in the development of the differential equations that might be incorporated in a model, consider the simple case of ground water flow. The development begins with a mathematical statement of fluid mass conservation having the following form:

$$-\left[\frac{\delta(\rho q_x)}{\delta x}+\frac{\delta(\rho q_y)}{\delta y}+\frac{\delta(\rho q_z)}{\delta z}\right] \pm \rho W=\frac{\delta(\rho \epsilon)}{\delta t}, \tag{2.2}$$

where ρ is the fluid density, q is the specific discharge or Darcy velocity, ϵ is porosity, W is a source-sink rate term, t is time, and x, y, z represents the system of Cartesian coordinates.

This equation is developed by taking a small cube of porous medium and accounting for inflows and outflows, fluid mass storage, and sources or sinks (Freeze and Cherry, 1979).

What is more important than the mathematical complexity of this equation is what it means. The first three terms on the left-hand side of (2.2) incorporate all the processes contributing to fluid mass or dissolved mass movement, and the fourth accounts for sources and sinks. Every problem of flow that can be modeled is described by one or more equations of this form.

The continuity equation (2.3) is a general mathematical framework that needs to be refined by providing a more precise description of each of the processes involved. More specifically, this means expressing mass fluxes in terms of the driving force (i.e., head gradients) and source-sink terms in terms of specific rate equations for the particular processes. To illustrate these ideas, consider the development of the basic mass flux equations beginning with the simplest case, that of ground water flow.

Freeze and Cherry (1979) show how to simplify (2.2) by interpreting the right-hand side in terms of water released from storage because of a decline in head, and assuming fluid density is constant they divide all terms by ρ. With these modifications the statement of continuity can be rewritten as

$$-\left[\frac{\delta(q_x)}{\delta x}+\frac{\delta(q_y)}{\delta y}+\frac{\delta(q_z)}{\delta z}\right] \pm W=S_s \delta h / \delta t, \tag{2.3}$$

where S_s is specific storage and h is hydraulic head.

The continuity equation (2.3) needs to be refined by providing a more precise description of the flow process. More specifically, this means expressing mass fluxes in terms of the driving force (i.e., head gradients).

For ground water flow, the step involves replacing specific discharge by using the well-known Darcy equation. In its simplest form, the Darcy equation states that the flow of water through a porous medium with a unit cross-sectional area is related to the product of the hydraulic gradient and a constant of proportionality termed hydraulic conductivity. This latter parameter is related to the permeability of the medium. Mathematically, this important physical law can be written as

$$\begin{aligned} q_x &= -K_x \delta h/\delta x, \\ q_y &= -K_y \delta h/\delta y, \\ q_z &= -K_z \delta h/\delta z, \end{aligned} \tag{2.4}$$

where q_x is the specific discharge or Darcy velocity and K is the hydraulic conductivity with components in the x, y, and z directions that are aligned with the principal material property axes. The negative signs in (2.4) mean that water is flowing in the direction opposite to increasing hydraulic potentials.

Substitution of (2.4) in (2.3) provides one form of an equation suitable for modeling ground water flow, or

$$\frac{\delta}{\delta x}(K_x \frac{\delta h}{\delta x}) + \frac{\delta}{\delta y}(K_y \frac{\delta h}{\delta y}) + \frac{\delta}{\delta z}(K_z \frac{\delta h}{\delta z}) \pm W = S_s \delta h/\delta t. \tag{2.5}$$

In this equation, the source-sink term does not need further elaboration because for the case of ground water flow, it is a simple constant related to the pumping or injection rate per unit volume. Thus (2.5) in one form or another is the differential equation used to model ground water flow in response to a potential gradient and subject to the effects of pumping/injection.

In some cases, fluid properties such as density or viscosity vary significantly in time or space because of changes in temperature or chemical composition. When the system is nonhomogeneous, the relations among water levels, heads, pressures, and velocities are not straightforward. Calculations of flow rates and directions then require information on intrinsic permeability, density, and viscosity (rather than hydraulic conductivity) and fluid pressures and elevations (rather than hydraulic heads).

The same set of steps—(1) development of the continuity equation and (2) substitution of some form of the Darcy equation—can be used with every fluid flow problem. However, for multicomponent flow problems, one continuity equation is required for each flowing fluid. Tables 2.4 and 2.5 summarize the development of basic equations for unsaturated ground water flow and two-component liquid flow (organic liquid and water), respectively. The general steps are the same in both cases.

The solution to the unsaturated flow equation (2.6) (see Table

TABLE 2.4 Summary of the Steps Involved in Developing the Unsaturated Flow Equation

Write the continuity equation for water—an equation for the air phase is not required because the air phase is assumed to be immobile. Also no source term is assumed. Θ is the volumetric moisture content; Θ' is the degree of saturation.	$-\left[\frac{\partial(\rho q_x)}{\partial x} + \frac{\partial(\rho q_y)}{\partial y} + \frac{\partial(\rho q_z)}{\partial z}\right] = \frac{\partial(\rho\epsilon\Theta')}{\partial t}$
Make assumptions that let ρ be removed, i.e., constant density. Right-hand side has been expanded and small terms removed, with $\Theta = \epsilon\Theta'$.	$-\left[\frac{\partial q_x}{\partial x} + \frac{\partial q_y}{\partial y} + \frac{\partial q_z}{\partial z}\right] = \frac{\partial\Theta}{\partial t}$
Substitute the Darcy equation for unsaturated flow in which hydraulic conductivity K is a function of pressure head ψ.	$q_x = -K(\psi)\frac{\partial h}{\partial x}$ $q_y = -K(\psi)\frac{\partial h}{\partial y}$ $q_z = -K(\psi)\frac{\partial h}{\partial z}$
Substitute q_x, q_y, q_z into continuity equation.	$\frac{\partial}{\partial x}\left[K(\psi)\frac{\partial h}{\partial x}\right] + \frac{\partial}{\partial y}\left[K(\psi)\frac{\partial h}{\partial y}\right] + \frac{\partial}{\partial z}\left[K(\psi)\frac{\partial h}{\partial z}\right] = \frac{\partial\Theta}{\partial t}$
Rewrite in terms of pressure head ψ, where $h = \psi + z$ and $\frac{\partial\phi}{\partial t} = \frac{\partial\phi}{\partial\psi}\cdot\frac{\partial\psi}{\partial t}$ or $C(\psi)\frac{\partial\psi}{\partial t}$, where $C(\psi)$ is the specific moisture capacity.	$\frac{\partial}{\partial x}\left[K(\psi)\frac{\partial\psi}{\partial x}\right] + \frac{\partial}{\partial y}\left[K(\psi)\frac{\partial\psi}{\partial y}\right] + \frac{\partial}{\partial z}\left[K(\psi)\left(\frac{\partial\psi}{\partial z} + 1\right)\right] = C(\psi)\frac{\partial\psi}{\partial t}$ (2.6)

NOTE: Solution of the working equation requires knowledge of the characteristic curves $K(\psi)$ and $C(\psi)$.
SOURCE: From Freeze and Cherry, 1979.

TABLE 2.5 Summary of the Steps Involved in Developing Two-Component Flow Equations

Step	Equations	
Write the continuity equations for each fluid. Subscripts *w* and *n* denote water and contaminant, respectively. Right-hand side uses ϵS_w and ϵS_n to reflect the fact that the porosity contains immiscible fractions of both. (Note the constraints applied to the solution because of the interrelationship of relative permeabilities and saturations in multiphase systems.)	*Water* $-\left[\frac{\partial(\rho_w q_{wx})}{\partial x} + \frac{\partial(\rho_w q_{wy})}{\partial y} + \frac{\partial(\rho_w q_{wz})}{\partial z}\right] = \frac{\partial(\rho_w \epsilon S_w)}{\partial t}$ *Contaminant* $-\left[\frac{\partial(\rho_n q_{nx})}{\partial x} + \frac{\partial(\rho_n q_{ny})}{\partial y} + \frac{\partial(\rho_n q_{nz})}{\partial z}\right] = \frac{\partial(\rho_n \epsilon S_n)}{\partial t}$	
Substitute the pressure forms of Darcy's equation shown, where hydraulic conductivity is expressed in terms of permeability (*k*), relative permeability (k_r), and dynamic viscosity (μ).	*Water* $q_{wx} = -\frac{kk_{rw}}{\mu_w}\left(\frac{\partial p_w}{\partial x}\right)$ $q_{wy} = -\frac{kk_{rw}}{\mu_w}\left(\frac{\partial p_w}{\partial y}\right)$ $q_{wz} = -\frac{kk_{rw}}{\mu_w}\left(\frac{\partial p_w}{\partial z} + \rho_w g\right)$	*Contaminant* $q_{nx} = -\frac{kk_{rn}}{\mu_n}\left(\frac{\partial p_n}{\partial x}\right)$ $q_{ny} = -\frac{kk_{rn}}{\mu_n}\left(\frac{\partial p_n}{\partial y}\right)$ $q_{nz} = -\frac{kk_{rn}}{\mu_n}\left(\frac{\partial p_n}{\partial z} + \rho_n g\right)$

TABLE 2.5 *Continued*

Arrive at working equations to be solved for pressure and saturation of both water and contaminant phases.	*Water* $$\frac{\partial}{\partial x}\left[\frac{k\rho_w k_{rw}}{\mu_w}\frac{\partial p_w}{\partial x}\right] + \frac{\partial}{\partial y}\left[\frac{k\rho_w k_{rw}}{\mu_w}\frac{\partial p_w}{\partial y}\right] + \frac{\partial}{\partial z}\left[\frac{k\rho_w k_{rw}}{\mu_w}\left(\frac{\partial p_w}{\partial z} + \rho_w g\right)\right] = \partial\left(\frac{\rho_w \epsilon S_w}{\partial t}\right) \quad (2.7)$$ *Contaminant* $$\frac{\partial}{\partial x}\left[\frac{k\rho_n k_{rn}}{\mu_n}\frac{\partial p_n}{\partial x}\right] + \frac{\partial}{\partial y}\left[\frac{k\rho_n k_{rn}}{\mu_n}\frac{\partial p_n}{\partial y}\right] + \frac{\partial}{\partial z}\mathrm{r}\left[\frac{k\rho_n k_{rn}}{\mu_n}\left(\frac{\partial p_n}{\partial z} + \rho_n g\right)\right] = \partial\left(\frac{\rho_n \epsilon S_n}{\partial t}\right) \quad (2.8)$$

2.4) provides the pressure head (ψ) at selected points in the unsaturated zone as a function of time. In Table 2.5, the two equations describing the problem of multicomponent flow, (2.7) and (2.8), can be used together with other relationships to predict pressure and saturation distributions for the two fluids as a function of time and space. Clearly, these equations do not describe all of the complexity of multiple-fluid systems. For example, Abriola and Pinder (1985a,b) describe a comprehensive approach to modeling the migration of a chemical contaminant in the nonaqueous phase, in the gas phase, and in the water phase as a soluble component.

The differential equations describing the transport of mass dissolved in ground water also are developed from conservation statements. However, the processes involved are quite different. Let the flux of a particular dissolved constituent into and out of a volume element of porous medium be represented by J. The change in notation to J simply reflects that dissolved mass rather than fluid mass is being transported. The continuity equation has the following form:

$$-\left[\frac{\delta(J)}{\delta x}+\frac{\delta(J)}{\delta y}+\frac{\delta(J)}{\delta z}\right]\pm r=\frac{\delta m}{\delta t}, \tag{2.9}$$

where r is a source-sink term accounting for mass lost or gained within the volume element and m is the mass per unit volume. The mass of a particular contaminant dissolved in a unit volume is the concentration (C; mass per unit volume of solution) multiplied by the porosity (ϵ), and so (2.9) becomes

$$-\left[\frac{\delta(J_x)}{\delta x}+\frac{\delta(J_y)}{\delta y}+\frac{\delta(J_z)}{\delta z}\right]\pm r=\frac{\delta(\epsilon C)}{\delta t}. \tag{2.10}$$

As before, the mass flux and the source terms on the left side of (2.10) are replaced by more detailed expressions describing the processes. The mass transport of a dissolved species is controlled by three processes: advection, diffusion, and dispersion. Mathematically, the advective flux of a contaminant is described by the following equation:

$$J_x = v_x C\epsilon, \tag{2.11}$$

where v_x is the mean ground water velocity or seepage velocity in the x direction. Dispersive mass fluxes are commonly assumed to be driven by concentration gradients, the so-called diffusive model of dispersion, and as such are described by Fick's law:

$$J_x = -\epsilon D_x \delta C / \delta x, \tag{2.12}$$

where D_x is the dispersion coefficient in the x direction. The dispersion coefficient includes both the contributions from true molecular diffusion and hydrodynamic mixing (Bear, 1972, 1979; Freeze and Cherry, 1979). A detailed discussion of the mathematical formulation of dispersion concepts is presented by Bear (1972, 1979) and Anderson (1979, 1984).

The combined effects of advection and dispersion can be accounted for simply by adding (2.11) and (2.12) to give

$$\begin{aligned} J_x &= -\epsilon D_x \delta C / \delta x + v_x C \epsilon, \\ J_y &= -\epsilon D_y \delta C / \delta y + v_y C \epsilon, \\ J_z &= -\epsilon D_z \delta C / \delta z + v_z C \epsilon. \end{aligned} \tag{2.13}$$

Substituting (2.13) into (2.10) provides a useful form of the advection-dispersion equation. Because this equation in three dimensions is so unwieldy, the following one-dimensional version,

$$-\left[\frac{\delta}{\delta x}\left(-\epsilon D_x \frac{\delta C}{\delta x} + v_x C \epsilon\right)\right] \pm r = \frac{\delta(\epsilon C)}{\delta t}, \tag{2.14}$$

is used in the committee's discussion of chemical and microbiological attenuation mechanisms. Assuming porosity to be constant in space and time and D_x constant in space, (2.14) can be rewritten as

$$D_x \frac{\delta^2 C}{\delta x^2} - v_x \frac{\delta C}{\delta x} \pm \frac{r}{\epsilon} = \frac{\delta C}{\delta t}. \tag{2.15}$$

Equation (2.15) is a common form of the advection-dispersion reaction equation that is just about ready to use except for the source-sink term. It is at this point that any of the reactions of interest (e.g., radioactive decay or ion exchange) have to be specified in detail. For example, examine the case of a first-order reaction describing either radioactive decay or hydrolysis:

$$r = d(\epsilon C)/dt = -\lambda \epsilon C, \tag{2.16}$$

where λ is the decay constant, related to the half-life for decay. All that is required to come up with one form of a simplified contaminant transport equation is to substitute (2.16) into (2.15) to give

$$D_x \frac{\delta^2 C}{\delta x^2} - v_x \frac{\delta C}{\delta x} - \lambda C = \frac{\delta C}{\delta t}. \tag{2.17}$$

The case of mass transport accompanied by sorption that is described in terms of a simple linear isotherm can be developed in the same way. Again, an expression for the source term has to be developed by starting with the isotherm

$$S = K_d C, \tag{2.18}$$

where S is the quantity of mass sorbed on the surface and K_d is the distribution coefficient. The appropriate rate expression is the product of the bulk density of the medium (ρ_{aq}) and the time derivative of (2.18), or

$$-r = \rho_{\mathrm{aq}} \frac{\delta S}{\delta t} = \rho_{\mathrm{aq}} K_d \frac{\delta C}{\delta t}. \tag{2.19}$$

Substituting (2.19) into (2.15) provides the governing mass transport equation, or

$$D_x \frac{\delta^2 C}{\delta x^2} - v_x \frac{\delta C}{\delta x} - \rho_{\mathrm{aq}} \frac{K_d}{\epsilon} \frac{\delta C}{\delta t} = \frac{\delta C}{\delta t}.$$

Rearranging terms yields

$$D_x \frac{\delta^2 C}{\delta x^2} - v_x \frac{\delta C}{\delta x} = \frac{\delta C}{\delta t}(1 + \frac{\rho_{\mathrm{aq}} K_d}{\epsilon}), \tag{2.20}$$

where the quantity in parentheses on the right side of (2.20) is a constant known as the retardation factor (R_f).

It is beyond the scope of this overview to examine how the many different chemical and biological processes are specifically developed into transport equations like (2.17). The process is essentially the same as just described. What makes formulating equations involving complex reactions somewhat difficult is that it is necessary to write an equation like (2.17) for each contaminant that is being transported and participating in the reaction. For example, a relatively complete description of the aerobic biodegradation of an organic contaminant requires an equation describing the transport of the organic contaminant and oxygen as well as a growth model describing how the mass of the microbial population changes with time (Molz et al., 1986). Each additional equation increases the data requirements. An additional problem is that some of these more complex reactions are not well understood, which adds uncertainty to the mathematical models that represent them.

Boundary and Initial Conditions and Parameter Values

The equations of flow and transport in themselves are general statements of how fluids or dissolved mass in a system should behave as a consequence of controlling processes. However, before one proceeds to actually solve an equation, information is needed about the system. There are essentially three features that need to be described: (1) the size and shape of the region of interest, (2) the boundary and initial conditions for that region, and (3) the physical and chemical properties that describe and control the processes in the system.

To illustrate these ideas, consider a problem involving the application of a steady-state ground water flow model to the field (Figure 2.12). At the particular site shown in Figure 2.12, glacial till overlies sandstone and shale bedrock. The first problem to address is definition of the region of interest. In other words, the lateral and vertical dimensions of the area to be modeled must be determined. The thick lines on Figure 2.12a define the area selected. The bottom boundary is assumed to coincide with the top of the Bearpaw shale, the upper boundary is the water table, and the lateral boundaries are vertical lines drawn at a major topographic high and a major topographic low.

When a region has been defined, it is implicitly assumed that the rest of the geologic system can be ignored (Figure 2.12b). However, the simulation has to account for the effects of conditions outside of the region being modeled. This job is handled by the boundary conditions applied on four sides. The boundary conditions are what make it possible to isolate a specific region of interest for detailed study.

There are three commonly used boundary conditions: (1) specified value, (2) specified flux, and (3) value-dependent flux (Mercer and Faust, 1981). These are briefly described in Table 2.6. It is important to realize that every differential equation included in a model requires a unique set of boundary conditions.

For the problem in Figure 2.12, the bottom and lateral side boundaries are assumed to be no-flow boundaries. The choice of a no-flow boundary on the bottom can be justified by geologic arguments; that is, the hydraulic conductivity of the shale is several orders of magnitude smaller than overlying units. The side boundaries are no-flow by virtue of the assumed symmetry of flow on either side (the boundaries represent flowlines). By intentionally placing these boundaries at a topographic high and a topographic low, the

TABLE 2.6 Typical Boundary Conditions for Ground Water Flow and Transport Problems

Type	Description
Specified value	Values of head, concentration, or temperature are specified along the boundary. (In mathematical terms, this is known as the Dirichlet condition.)
Specified flux	Flow rate of water, contaminant mass, or energy is specified along the boundary and equated to the normal derivative. For example, the volumetric flow rate per unit area for water in an isotropic medium is given by $$q_n = -K\frac{\partial h}{\partial n},$$ where the subscript n refers to the direction normal (perpendicular) to the boundary. (A medium that is isotropic with respect to hydraulic conductivity is equally permeable in all directions.) A no-flow (impermeable) boundary is a special case of this type in which $q_n = 0$. (When the derivative is specified on the boundary, it is called a Neuman condition.)
Value-dependent flux	The flow rate is related to both the normal derivative and the value. For example, the volumetric flow rate per unit area of water is related to the normal derivative of head and to head itself by $$-K\frac{\partial h}{\partial n} = q_n(h_b),$$ where q_n is some function that describes the boundary flow rate given the head at the boundary (h_b).

SOURCE: Mercer and Faust, 1981.

flowlines will generally parallel these boundaries and make them no-flow boundaries. The actual decision of what the modeled region would be like in this example was made in part to provide a simple set of boundary conditions. Moving the lateral boundaries could create a much more complex set of boundary conditions. However, the modeler does have the option of placing the boundaries anywhere. Note that similar boundary conditions (e.g., specified concentrations or mass fluxes) are required to solve each mass transport equation applied to the domain.

The water table boundary is an example of a specified-value

condition. Values of hydraulic head are assumed to be known at all points along the water table. In some problems, flow across the top boundary is represented by recharge and discharge fluxes, with the configuration of the water table actually determined as part of the simulation.

The last information needed about the system is the value of parameters controlling the various flow and transport processes. Assuming the example system to be at steady state, the only parameters necessary are values of the hydraulic conductivity (K_x and K_y) for each geologic unit (Figure 2.12b) and injection or withdrawal rates for sources or sinks. At this point, the flow equation is ready to be solved to provide the unknown hydraulic head at points within the region.

If the flow problem in Figure 2.12 was transient, it would be necessary to provide the initial conditions, or, in other words, the distribution of hydraulic head in the region, at the start of the simulation. In addition, values for specific storage must also be specified. With this information, it would be possible to simulate the changing conditions of hydraulic head not only as a function of space but also as a function of time.

The discussion of the three preparatory steps to modeling is related particularly to a simple problem of ground water flow. However, the same steps are followed for multicomponent flow and dissolved contaminant transport. All that change are the type and number of parameters because of the type and number of processes involved. For example, to model the transport of a single dissolved contaminant that may degrade in a first-order kinetic reaction requires values of velocity (v_x, v_y) everywhere in the domain of interest, longitudinal and transverse dispersivities, and the decay rate constant for the reaction. In many instances, it will be necessary to run a flow model to provide the necessary description of the velocity field. As indicated previously, several transport equations may be necessary, depending on the complexity of the degradation reaction, e.g., its dependence upon hydrogen, oxygen, or other substrates.

Solving Flow and Transport Equations

There are two basic ways to solve the flow and transport equations. The analytical methods embody classical mathematical approaches that have been used for more than 100 years to deal with differential equations. The numerical approaches have also existed for many years but were not fully exploited until the development of computers to solve approximate forms of the governing equations. The greatest strength of the analytical methods lies in their capability in many cases to produce exact solutions to a flow or transport problem in terms of the controlling parameters. Being able to establish the functional form of the solution, and the interrelationships among parameters, provides a great deal of physical insight into how the processes control flow and transport. Another useful way in which the analytical solutions are used is to provide a check on the accuracy of numerical models, which can be subject to a variety of different errors.

In terms of their usefulness in solving practical problems, the numerical approaches are superior to the analytical methods because the user can let the controlling parameters vary in space and time. This feature enables detailed replications of the complex geologic and hydrologic conditions that exist in nature. Analytical methods have a role to play in field applications (e.g., theory of well hydraulics), but, in general, they are appropriate only for a narrow range of simple problems. Practical problems involving the flow of more than one fluid or contaminant are sufficiently complex that only numerical approaches are suitable.

Nearly all the numerical procedures involve replacing the continuous form of the governing differential equation by a finite number of algebraic equations. To develop these equations, it is necessary to subdivide the region into pieces. For the flow example discussed previously (Figure 2.12), the region can be subdivided by using rectangles (Figure 2.13). Other geometric shapes (e.g., triangles and quadrilaterals) are also used, depending on the solution technique. For transient problems, it is also necessary to subdivide the total simulation period into a number of smaller time steps.

For the example problem, values of hydraulic head are calculated at the nodes, located at the center of cells, with one algebraic equation written for each node. Hydraulic conductivity values are supplied for each rectangular cell. This flexibility in assigning parameter values helps create in the model a distribution of geohydrologic properties that closely approximates that observed in the field.

A variety of analytical and numerical solutions have been developed for use in ground water applications, and a comprehensive discussion of each would require a modeling textbook. However, a summary of techniques that are commonly applied to the solution of various flow and mass transport problems is provided in Table

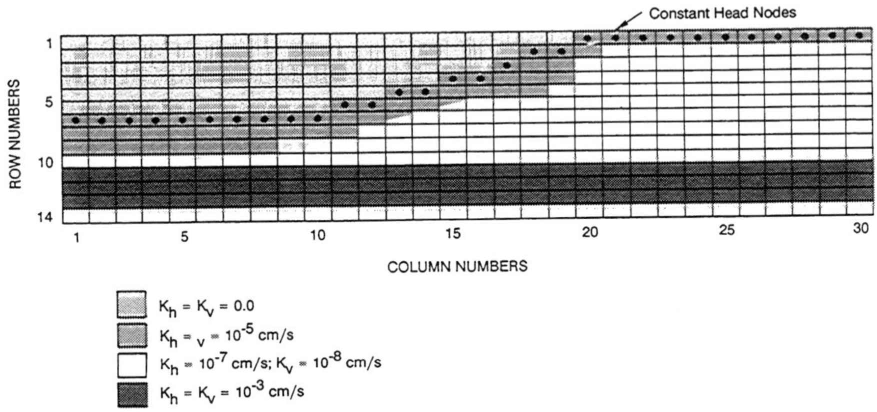

FIGURE 2.13 Model system from the previous figure subdivided by a rectangular grid system. Nodes are defined in the center of each grid cell. By assigning $K_h = K_v = 0.0$ in the area at the top left, it is effectively excluded from the calculation.

2.7. Included are a brief description of the methods and a few key references that can be used to obtain more detailed information.

The last topic that needs to be addressed in this section involves the mathematical techniques for solving matrix equations. In most models, the major computational effort comes in solving the system of model equations. In general, there are two basic methods. In one approach the entire system of equations is solved simultaneously with direct methods, providing a solution that is exact, except for machine round-off error. In the second approach, iterative methods obtain a solution by a process of successive approximation, which involves making an initial guess at the matrix solution and then improving this guess by some iterative process until an error criterion is satisfied.

Direct methods have two main disadvantages. The first is that a computer may not be able to store the large matrices or solve the system in a reasonable time when the number of nodes is large. Sometimes this problem can be dealt with to some extent by using sparse matrix solvers and various node-numbering schemes. The second problem with direct methods is the round-off errors. Because many arithmetic operations are performed, round-off errors can accumulate and significantly influence results for certain types of matrices.

Iterative schemes avoid the need for storing large matrices. This

TABLE 2.7 A Summary of the Common Solution Techniques for Problems of Fluid Flow and Dissolved Mass Transport

Problem	Solution Technique	Description/Comments on the Method	Key References
Ground water flow	Analytical	Involves the use of classical mathematical techniques for solving differential equations. Widely used for more than 35 years in ground water resource evaluation.	Hantush (1964) Jacob (1940) Kruseman and de Ridder (1983) Lohman (1979) Theis (1935) Walton (1970)
	Finite difference	Uses differential equations to approximate derivatives, resulting in a series of algebraic equations. Widely used since the 1960s with few limitations.	Mercer and Faust (1981) Remson et al. (1971) Wang and Anderson (1982)
	Finite element	Creates an integral form of the differential equation; again discretization provides a system of linear algebraic equations. Widely used with few limitations. The ability to use a variety of element shapes is helpful in subdividing irregularly shaped aquifer or geologic units.	Huyakorn and Pinder (1983) Mercer and Faust (1981) Pinder and Gray (1977) Zienkiewicz (1977)
	Boundary elements or boundary integral methods	Creates integral form of the governing flow equation relying on boundary rather than areal integrals. By working with the boundaries of aquifers or units this method avoids internal discretization, and thus a small number of large elements can be used instead of the finite-element method.	DeMarsily (1986) Liggett and Liu (1983)

TABLE 2.7 *Continued*

Problem	Solution Technique	Description/Comments on the Method	Key References
Multiphase flow Unsaturated zone	Analytical	See above.	Lappala (1980) Nielsen et al. (1986) Philip (1955, 1957)
	Finite difference	See above; sharp changes in parameters at wetting front require consideration in grid design. Nonlinear problem because hydraulic conductivity is a function of pressure head.	Hanks et al. (1969) Jeppson (1974) Narasimhan et al. (1978) Neuman (1972) Neuman and Narasimhan (1977) Pruess (1980) Reisenauer (1963)
	Finite element	See above.	Yeh and Ward (1980)
Two-fluid flow	Finite difference	See above; solves flow equation for each fluid.	Abriola and Pinder (1985a,b) Baehr and Corapcioglu (1987) Corapcioglu and Baehr (1987) Faust (1985)
	Finite element	See above.	Lenhard and Parker (1987) Osborne and Sykes (1986) Parker and Lenhard (1987)
Dissolved contaminant transport	Analytical	See above; a variety of different solutions exists for contaminant transport in one, two, and three dimensions.	Cleary and Ungs (1978) Domenico and Robbins (1985) Javandel et al. (1984) Ogata (1970) van Genuchten and Alves (1982)
	Finite difference	See above; in advection-dominated problems, numerical dispersion and oscillations can develop in solution. Case of multicomponent transport with reaction requires special consideration.	Reeves et al. (1986a,b,c) Welch et al. (1966)

Dissolved contaminant transport *(continued)*	Finite element	See above; same problems as finite difference.	Huyakorn and Pinder (1983) Voss (1984) Yeh and Ward (1981)
	Method of characteristics	Breaks the advection-dispersion equation into two parts, one accounting for advection and the other accounting for dispersion. Requires the transport of reference particles.	Bredehoeft and Pinder (1973) Konikow and Bredehoeft (1978) Reddell and Sunada (1970)
	Random walk methods	One of the few techniques not involving a solution of the advection-dispersion equation. Simulates the migration of contaminants by moving a set of reference particles. Generally provides an approximate solution.	Ahlstrom et al. (1977) Prickett et al. (1981) Schwartz and Crowe (1980)

feature makes them attractive for solving problems with many unknowns. Numerous schemes have been developed; a few of the more commonly used ones include successive overrelaxation methods (Varga, 1962), the alternating-direction implicit procedure (Douglas and Rachford, 1956), the iterative alternating-direction implicit procedure (Wachpress and Habetler, 1960), and the strongly implicit procedure (Stone, 1968). Because operations are performed many times, iterative methods also suffer from potential round-off errors.

The efficiency of iterative methods depends on an initial estimate of the solution. This makes the iterative approach less desirable for solving steady-state problems (Narasimhan et al., 1978). To speed up the iterative process, relaxation and acceleration factors are used. Unfortunately, the definition of best values for these factors commonly is problem dependent. In addition, iterative approaches require that an error tolerance or convergence criterion be specified to stop the iterative process. This, too, may be problem dependent. All of these parameters must be specified by the model user.

According to Narasimhan et al. (1977) and Neuman and Narasimhan (1977), perhaps the greatest limitation of the iterative schemes is the requirement that the matrix be well conditioned. An ill-conditioned matrix can drastically affect the rate of convergence or even prevent convergence. An example of an ill-conditioned matrix is one in which the main diagonal terms are much smaller than other terms in the matrix.

More recently, a semi-iterative method has gained popularity (Gresho, 1986). This method, or class of methods known as conjugate gradient methods, was first described by Hestenes and Stiefel (1952). It is widely used to solve linear algebraic equations where the coefficient matrix is sparse and square (Concus et al., 1976). One advantage of the conjugate gradient method is that it does not require the use or specification of iteration parameters, thereby eliminating this partly subjective procedure (Manteuffel et al., 1983). Kuiper (1987) compared the efficiency of 17 different iterative methods for the solution of the nonlinear three-dimensional ground water flow equation. He concluded that, in general, the conjugate gradient methods did the best.

Numerical methods, by their very nature, yield approximate solutions to the governing partial differential equations. The accuracy of the solution can be significantly affected by the choice of numerical parameters, such as the size of the spatial discretization grid and the length of time steps. Those using ground water models and those making management decisions based on model results should always be aware that trade-offs between accuracy and cost will always have to be made. If the grid size or time steps are too coarse for a given problem, it is possible to generate a numerical solution that converges on an answer that has an excellent mass balance but is still inaccurate. Furthermore, if iteration parameters are not properly specified, the solution may not converge. It is hoped that this will show up as a mass balance error, which will be noted by the user. This indicates, however, the importance of a mass balance in numerical models.

REFERENCES

Abriola, L. M., and G. F. Pinder. 1985a. A multiphase approach to the modeling of porous media contamination by organic compounds, 1. Equation development. Water Resources Research 21(1), 11–18.

Abriola, L. M., and G. F. Pinder. 1985b. A multiphase approach to the modeling of porous media contamination by organic compounds, 2. Numerical simulation. Water Resources Research 21(1), 19–26.

Ahlstrom, S. W., H. P. Foote, R. C. Arnett, C. R. Cole, and R. J. Serne. 1977. Multicomponent Mass Transport Model: Theory and Numerical Implementation (discrete-parcel-random-walk version). BNWL-2127. Battelle Northwest Laboratories, Richland, Wash.

Alexander, M. 1985. Biodegradation of organic chemicals. Environmental Science and Technology 19, 106.

Anderson, M. P. 1979. Using models to simulate the movement of contaminants through ground water flow systems. Critical Reviews in Environmental Control 9(2), 97–156.

Anderson, M. P. 1984. Movement of contaminants in groundwater: Groundwater transport-advection and dispersion. Pp. 37–45 in Groundwater Contamination. Studies in Geophysics. National Academy Press, Washington, D.C.

Atlas, R. M. 1981. Microbial degradation of petroleum hydrocarbons: An environmental perspective. Microbiological Reviews 45, 180.

Baehr, A. L., and M. Y. Corapcioglu. 1987. A compositional multiphase model for groundwater contamination by petroleum products, 2. Numerical solution. Water Resources Research 23(1), 201–214.

Bates, J. K., and W. B. Seefeldt. 1987. Scientific Basis for Nuclear Waste Management X. Materials Research Society, Symposium Proceedings, Vol. 84, 829 pp.

Bear, J. 1972. Dynamics of Fluids in Porous Media. Elsevier, New York, 764 pp.

Bear, J. 1979. Hydraulics of Groundwater. McGraw-Hill, New York, 569 pp.

Bredehoeft, J. D., and G. F. Pinder. 1973. Mass transport in flowing groundwater. Water Resources Research 9, 194–210.

Cherry, J. A., R. W. Gillham, and J. F. Pickens. 1975. Contaminant hydrogeology: Part 1, Physical processes. Geoscience Canada 2(2), 76–84.

Cleary, R. W., and M. J. Ungs. 1978. Groundwater Pollution and Hydrology, Mathematical Models and Computer Programs. Rep. 78-WR-15, Water Resources Program, Princeton University, Princeton, N.J.

Concus, P., G. Golub, and D. O'Leary. 1976. Sparse Matrix Computations. Academic Press, New York, pp. 309–332.

Corapcioglu, M. Y., and A. L. Baehr. 1987. A compositional multiphase model for groundwater contamination by petroleum products, 1. Theoretical considerations. Water Resources Research 23(1), 191–200.

DeMarsily, G. 1986. Quantitative Hydrogeology. Academic Press, Orlando, Fla., 440 pp.

Domenico, P. A., and G. A. Robbins. 1985. A new method of contaminant plume analysis. Ground Water 23(4), 476–485.

Douglas, J., Jr., and H. H. Rachford, Jr. 1956. On the numerical solution of heat conduction problems in two and three space variables. Transactions of the American Mathematics Society 82, 421–439.

Faust, C. R. 1985. Transport of immiscible fluids within and below the unsaturated zone: A numerical model. Water Resources Research 21(4), 587–596.

Freeze, R. A. 1969a. Regional ground water flow—Old Wives Lake drainage basin, Saskatchewan. Inland Waters Branch, Department of Energy, Mines, and Resources, Canada, Scientific Series, No. 5, 245 pp.

Freeze, R. A. 1969b. Theoretical analysis of regional ground water flow. Inland Waters Branch, Department of Energy, Mines, and Resources, Canada, Scientific Series, No. 3.

Freeze, R. A. 1971a. Influence of the unsaturated flow domain on seepage through earth dams. Water Resources Research 7(4), 929–941.

Freeze, R. A. 1971b. Three-dimensional, transient, saturated-unsaturated flow in a groundwater basin. Water Resources Research 7(2), 347–366.

Freeze, R. A., and J. A. Cherry. 1979. Ground Water. Prentice-Hall, Englewood Cliffs, N.J., 604 pp.

Freyberg, D. L. 1986. A natural gradient experiment on solute transport in a sand aquifer: II. Spatial moments and the advection and dispersion of non-reactive tracers. Water Resources Research 22(13), 2031–2046.

Fried, J. J. 1975. Ground Water Pollution: Theory, Methodology, Modeling, and Practical Rules. Elsevier, Amsterdam, 330 pp.

Frind, E. O. 1987. Simulation of ground water contamination in three dimensions. Pp. 749–763 in Proceedings of Solving Ground Water Problems with Models. National Water Well Association, Denver, Colo.

Galloway, W. E., and D. K. Hobday. 1983. Terrigenous Clastic Depositional Systems. Springer-Verlag, New York, 423 pp.

Greenkorn, R. A. 1983. Flow Phenomena in Porous Media: Fundamentals and Applications in Petroleum, Water and Food Production. Marcel Dekker, New York, 550 pp.

Gresho, P. M. 1986. Time integration and conjugate gradient methods for the incompressible Navier-Stokes equations. Pp. 3–27 in Finite Elements in Water Resources, Proceedings of the 6th International Conference, Lisbon, Portugal. Springer, Berlin.

Hanks, R. J., A. Klute, and E. Bresler. 1969. A numeric method for estimating infiltration, redistribution, drainage, and evaporation of water from soil. Water Resources Research 5, 1064–1069.

Hantush, M. S. 1964. Hydraulics of wells. Advances in Hydroscience 1, 281–432.

Hestenes, M., and E. Stiefel. 1952. Methods of conjugate gradients for solving linear systems. Journal of Research of the National Bureau of Standards 49(6), 409–436.

Huyakorn, P. S., and G. F. Pinder. 1983. Computational methods in subsurface flow. Academic Press, New York, 473 pp.
Jackson, R. E., R. J. Patterson, B. W. Graham, J. Bahr, D. Belanger, J. Lockwood, and M. Priddle. 1985. Contaminant Hydrogeology of Toxic Organic Chemicals at a Disposal Site, Gloucester, Ontario. 1. Chemical Concepts and Site Assessment. NHRI Paper No. 23, IWD Scientific Series No. 141, National Hydrology Research Institute, Inland Waters Directorate, Environment Canada, Ottawa, Canada, 114 pp.
Jacob, C. E. 1940. On the flow of water in an elastic artesian aquifer. Transactions, American Geophysical Union 2, 574–586.
Javandel, I., C. Doughty, and C.-F. Tsang. 1984. Groundwater Transport: Handbook of Mathematical Models. Water Resources Monograph 10, American Geophysical Union, Washington, D.C., 228 pp.
Jeppson, R. W. 1974. Axisymmetric infiltration in soils—Numerical techniques of solution. Journal of Hydrology 23, 111–130.
Konikow, L. F., and J. D. Bredehoeft. 1978. Computer model of two-dimensional solute transport and dispersion in ground water. Technical Water Resources Inventory, Book 7, Chap. C2. U.S. Geological Survey, Reston, Va., 90 pp.
Kruseman, G. P., and N. A. de Ridder. 1983. Analysis and Evaluation of Pumping Test Data. Bulletin 11. International Institute for Land Reclamation and Improvement/ILRI, P.O. Box 45, G700 AA Wageningen, The Netherlands.
Kuiper, L. K. 1987. A comparison of iterative methods as applied to the solution of the nonlinear three-dimensional groundwater flow equation. SIAM Journal of Scientific and Statistical Computing 8(4), 521–528.
Lappala, E. G. 1980. Modeling of water and solute transport under variably saturated conditions: State of the art. Prepared for Proceedings of the Interagency Workshop on Radioactive Waste Modeling, December 2–4, 1980, Denver, Colo.
Lenhard, R. J., and J. C. Parker. 1987. A model for hysteretic constitutive relations governing multiphase flow, 2. Permeability-saturation relations. Water Resources Research 23(12), 2197–2206.
Liggett, J. A., and P. L-F. Liu. 1983. The Boundary Integral Equation Method for Porous Media Flow. George Allen and Unevin, London.
Lindberg, R. D., and D. D. Runnells. 1984. Ground water redox reactions: An analysis of equilibrium state applied to Eh measurements and geochemical modeling. Science 225, 925–927.
Lohman, S. W. 1979. Ground-water hydraulics. U.S. Geological Survey Professional Paper 708. U.S. Government Printing Office, Washington, D.C.
Mackay, D. M., D. L. Freyberg, P. V. Roberts, and J. A. Cherry. 1986. A natural gradient experiment on solute transport in a sand aquifer, 1. Approach and overview of plume movement. Water Resources Research 22(13), 2017, 2029.
Manteuffel, T. A., D. B. Grove, and L. F. Konikow. 1983. Application of the conjugate-gradient method to ground-water models. Rep. 83-4009, Water Resources Investigations, U.S. Geological Survey, 24 pp.
Mercer, J. W., and C. R. Faust. 1981. Ground-Water Modeling. National Water Well Association, Worthington, Ohio, 60 pp.

Molz, F. J., M. A. Widdowson, and L. D. Benefield. 1986. Simulation of microbial growth dynamics coupled to nutrient and oxygen transport in porous media. Water Resources Research 22, 1207–1216.

Morel, F. M. M. 1983. Principles of Aquatic Chemistry. Wiley, New York, 446 pp.

Narasimhan, T. N., S. P. Neuman, and A. L. Edwards. 1977. Mixed explicit-implicit iterative finite element scheme for diffusion-type problems: II. Solution in strategy and examples. International Journal for Numerical Methods in Engineering 11, 325–344.

Narasimhan, T. N., P. A. Witherspoon, and A. L. Edwards. 1978. Numerical model for saturated-unsaturated flow in deformable porous media, 2. The algorithm. Water Resources Research 14(2), 255–261.

Neuman, S. P. 1972. Finite element computer programs for flow in saturated-unsaturated porous media. Second Annual Report. Project No. A10-SWC-77, Hydraulic Engineering Laboratory, Technion, Haifa, Israel, p. 87.

Neuman, S. P., and T. N. Narasimhan. 1977. Mixed explicit-implicit iterative finite element scheme for diffusion-type problems: I. Theory. International Journal for Numerical Methods in Engineering 11, 309–323.

Nielsen, D. R., M. Th. Van Genuchten, and J. W. Biggar. 1986. Water flow and solute transport processes in the unsaturated zone. Water Resources Research 22(9), 89S–108S.

Ogata, A. 1970. Theory of dispersion in a granular medium. U.S. Geological Survey Progress Paper 411-I.

Osborne, M., and J. Sykes. 1986. Numerical modeling of immiscible organic transport at the Hyde Park landfill. Water Resources Research 22(1), 25–33.

Parker, J. C., and R. J. Lenhard. 1987. A model for hysteretic constitutive relations governing multiphase flow, 1. Saturation-pressure relations. Water Resources Research 23(12), 2187–2196.

Philip, J. R. 1955. Numerical solution of equations of the diffusion type with diffusivity concentration dependent. Transactions of the Faraday Society 51, 885–892.

Phillip, J. R. 1957. The theory of infiltration: 1. The infiltration equation and its solution. Soil Science 83, 345–357.

Pinder, G. F., and W. G. Gray. 1977. Finite Element Simulation in Surface and Subsurface Hydrology. Academic Press, New York, 295 pp.

Prickett, T. A., T. G. Naymik, and C. G. Lounquist. 1981. A Random-Walk Solute Transport Model for Selected Groundwater Quality Evaluations. Bulletin 65, Illinois State Water Survey, Champaign, Ill., 103 pp.

Pruess, K., and R. C. Schroeder. 1980. SHAFT79, User's Manual. LBL-10861, Lawrence Berkeley Laboratory, University of California, Berkeley.

Reddell, D. L., and D. K. Sunada. 1970. Numerical simulation of dispersion in groundwater aquifer. Hydrol. Paper 41, Colorado State University, Fort Collins, 79 pp.

Reeves, M., D. S. Ward, P. A. Davis, and E. J. Bonano. 1986a. SWIFT II Self-Teaching Curriculum: Illustrative Problems for the Sandia Waste-Isolation Flow and Transport Model for Fractured Media. NUREG/CR-3925, SAND84-1586, Sandia National Laboratory, Albuquerque, N. Mex.

Reeves, M., D. S. Ward, N. D. Johns, and R. M. Cranwell. 1986b. Data Input Guide for SWIFT II, The Sandia Waste-Isolation Flow and Transport

Model for Fractured Media. NUREG/CR-3162, SAND83-0242, Sandia National Laboratories, Albuquerque, N. Mex.

Reeves, M., D. S. Ward, N. D. Johns, and R. M. Cranwell. 1986c. Theory and Implementation for SWIFT II, The Sandia Waste-Isolation Flow and Transport Model for Fractured Media. NUREG/CR-3328, SAND83-1159, Sandia National Laboratories, Albuquerque, N. Mex.

Reisenauer, A. E. 1963. Methods for solving problems of partially saturated steady flow in soils. Journal of Geophysical Research 68, 5725–5733.

Remson, I., G. M. Hornberger, and F. J. Molz. 1971. Numerical Methods in Subsurface Hydrology. Wiley, New York, 389 pp.

Rittmann, B. E., D. Jackson, and S. L. Storck. 1988. Potential for treatment of hazardous organic chemicals with biological processes. Pp. 15–94 in Biotreatment Systems, Vol. III, D. L. Wise, ed. CRC Press, Boca Raton, Fla.

Runnells, D. D. 1976. Wastewaters in the Vadose zone of arid regions: Geochemical interactions. Ground Water 14(6), 374–385.

Schwartz, F. W. 1975. On radioactive waste management: An analysis of the parameters controlling subsurface contaminant transport. Journal of Hydrology 27, 51–71.

Schwartz, F. W. 1977. Macroscopic dispersion in porous media: The controlling factors. Water Resources Research 13(4), 743–752.

Schwartz, F. W. 1984. Modeling of ground water flow and composition. Pp. 178–188 in Proceedings of First Canadian/American Conference in Hydrogeology, B. Hitchon and E. I. Wallich, eds. National Water Well Association, Banff, Alberta.

Schwartz, F. W., and A. S. Crowe. 1980. A deterministic-probabilistic model for contaminant transport. NUREG/CR-1609, Nuclear Regulatory Commission, 158 pp.

Schwille, F. 1984. Migration of organic fluids immiscible with water in the unsaturated zone. Pp. 27–48 in Pollutants in Porous Media: The Unsaturated Zone Between Soil Science and Groundwater, B. Yaron, G. Dagan, and J. Goldshmid, eds. Ecological Studies, Vol. 47. Springer-Verlag, Berlin.

Smith, L., and F. W. Schwartz. 1980. Mass transport, 1. A stochastic analysis of macroscopic dispersion. Water Resources Research 16(2), 303–313.

Stone, H. K. 1968. Iterative solution of implicit approximations of multidimensional partial differential equations. Society of Industrial and Applied Mathematics, Journal of Numerical Analysis, 5(3), 530–558.

Sudicky, E. A. 1986. A natural gradient experiment on solute transport in a sand aquifer: Spatial variability of hydraulic conductivity and its role in the dispersion process. Water Resources Research 22(13), 2069–2082.

Tennessee Valley Authority. 1985. A review of field-scale physical solute transport processes in saturated and unsaturated porous media. EPRI EA-4190, Project 2485-5, Electric Power Research Institute, Palo Alto, Calif.

Theis, C. V. 1935. The relationship between the lowering of the piezometric surface and the rate and duration of discharge of a well using groundwater storage. Transactions, American Geophysical Union 2, 519–524.

van Genuchten, M. T., and W. J. Alves. 1982. Analytical solutions of the one-dimensional convective-dispersive solute transport equation. Technical Bulletin 1661, U.S. Department of Agriculture, Washington, D.C., 149 pp.

Varga, R. S. 1962. Matrix Iterative Analysis. Prentice-Hall, Englewood Cliffs, N.J., 322 pp.

Voss, C. I. 1984. SUTRA—Saturated Unsaturated Transport—A finite-element simulation model for saturated-unsaturated fluid-density-dependent ground-water flow with energy transport or chemically-reactive single-species solute transport. Water Resources Investigations Rep. 84-4369, U.S. Geological Survey, Reston, Va., 409 pp.

Wachpress, E. L., and G. J. Habetler. 1960. An alternating-direction-implicit iteration technique. Journal of Society of Industrial and Applied Mathematics 8, 403–424.

Walton, W. C. 1970. Groundwater Resource Evaluation. McGraw-Hill, New York, 664 pp.

Wang, J. F., and M. P. Anderson. 1982. Introduction to Groundwater Modeling. Freeman, San Francisco, Calif., 237 pp.

Welch, J. E., F. H. Harlow, J. P. Shannon, and B. J. Daly. 1966. The MAC Method, A Computing Technique for Solving Viscous, Incompressible, Transient Fluid-Flow Problems Involving Free Surfaces. LA-3425, Los Alamos Scientific Laboratory of the University of California, Los Alamos, N. Mex.

Yeh, G. T., and D. S. Ward. 1980. FEMWATER: A Finite-Element Model of Water Flow Through Saturated-Unsaturated Porous Media. ORNL-5567, Oak Ridge National Laboratory, Oak Ridge, Tenn.

Yeh, G. T., and D. S. Ward. 1981. FEMWASTE: A Finite-Element Model of Waste Transport Through Saturated-Unsaturated Porous Media. ORNL-5601, Oak Ridge National Laboratory, Oak Ridge, Tenn.

Zienkiewicz, O. C. 1977. The Finite Element Method, 3rd ed. McGraw-Hill, London.

3

Flow Processes

INTRODUCTION

As suggested by the discussion in the previous chapter, most applications of ground water models to aid decisionmaking begin with the potential energy and flow of water alone. First of all, ground water flow models, with their focus on the prediction of head, volumes, and velocity of flow, can be important tools in the assessment and development of water resources. For example, predictions of the economic yield of an aquifer, or of the impacts of new or increased pumping on existing wells, or of ground water recharge below irrigated agriculture, all require an understanding and prediction of ground water head and flow. Second, ground water flow models are a crucial component of all analyses of contaminant transport because of the need to define the ground water velocity field. As noted in a number of sections of this report, advection with the flow field is often the dominant process controlling the direction, if not the rate, of transport. In the absence of significant density differences caused by contaminant concentration differences, the velocity field is independent of chemical and biological transport processes. Thus transport modeling studies usually begin with a prediction of the velocity field based on a ground water flow model.

Historically, the earliest ground water models were developed to predict head and volumetric flow in fully saturated, porous (nonfractured) geologic environments. Because of this relatively long history, saturated continuum flow models have been investigated extensively and are quite well understood in the context of a wide variety of problems. The modeling of flow in unsaturated, nonfractured settings has a shorter history and is largely dominated by problems of understanding and predicting infiltration from rainfall, irrigation, rivers, canals, and ponds. Our understanding of such models is less sophisticated than our understanding of saturated flow models. Least well developed, and indeed only in its infancy, is our understanding of recent attempts to model head and flow in both saturated and unsaturated fractured environments. For all three cases—saturated continuum flow, unsaturated continuum flow, and fracture flow—most of our understanding of flow modeling has been gained for problems requiring predictions of head and volumetric flow rates. The demands of contaminant transport prediction, in which the critical flow variable is velocity, are more challenging and have only recently become the focus of attention in ground water flow modeling.

Although ground water flow modeling is older and more advanced than ground water transport modeling, many issues and uncertainties remain in the application of flow models to decisionmaking problems, particularly those involving transport. This chapter summarizes the committee's sense of the state of the art of ground water flow modeling and of the issues related to the current and future use of flow models in decisionmaking.

SATURATED CONTINUUM FLOW

As mentioned in the introduction to this chapter, the earliest models of ground water were saturated continuum flow models. There are several reasons for this early interest in fully saturated flow. First, the dominant problems 30 years ago were problems of water resources development. Attention was focused on questions of available ground water resources and on the impacts of the installation of wells on these resources (and surface water resources). Variables of interest were head and volumetric flow rate. Relevant spatial scales were large—aquifers and aquifer systems. Ground water quality was assumed to be high, except in areas of saltwater intrusion.

A second impetus for the early focus on fully saturated flow was the relative ease with which the physical processes of greatest importance to water resources questions can be represented. Governing equations are linear or nearly so; assumptions of spatial homogeneity and temporal steady state are often justified; one- and two-dimensional formulations yield relevant, useful information; and relatively unsophisticated numerical approximation techniques are adequate. Put simply, the easiest types of ground water problems to model are fully saturated flow problems.

This ease is only relative, however. Fully saturated ground water flow modeling remains quite challenging. This is especially true for problems of contaminant transport below the water table. For such problems, the role of ground water flow modeling is to provide an estimate of the flow velocities. Head predictions are of little direct interest. Velocity estimates, however, are usually based on hydraulic head differences and therefore are much more sensitive to modeling errors than are estimates of hydraulic head alone. In addition, satisfactory predictions of transport often require that the velocity field be well predicted on fine spatial grids. The use of large-scale average velocities, which are usually very adequate for water supply problems, can place high demands on the dispersive component of a transport model, demands that we are only just beginning to understand (see Chapters 2 and 4). This need for high spatial resolution presents formidable challenges for data collection, parameter estimation, model formulation, numerical methods, and computational power and speed.

State of the Art

The physical processes controlling the flow of water through fully saturated porous rock or soil are well understood, both theoretically and experimentally. The mathematical statements of the fundamental physical laws governing general fluid motion—conservation of mass, momentum, and energy—which are collectively known as the Navier-Stokes equations, are universally accepted (White, 1974). More important, the simplifications of these equations, which lead to Darcy's law for fully saturated flow through porous media (equation [2.4]), have been investigated both in the laboratory and in theory. The Darcy equation is known to yield good predictions of head and flow under a wide range of conditions encountered in the subsurface (cf. Freeze and Cherry, 1979).

The conditions under which the Darcy equation is not adequate for prediction have been reasonably well delineated. Darcy's law is known to fail for high-velocity flows, which might occur in very porous gravel or boulder deposits, karst terrain, or the immediate vicinity of pumping wells (Bear, 1972, pp. 125–127). Darcy's law is suspected to fail for flow through extremely small pores under low-pressure head gradients (Freeze and Cherry, 1979, p. 72), conditions that might occur at great depth, for example, in the vicinity of potential radioactive waste repositories. Considerable controversy about this behavior exists. In neither case of failure is there adequate support for universally accepted alternative formulations (short of the Navier-Stokes equations), although a number of models have been proposed (Bear, 1972, pp. 176–184). Prediction uncertainty for these flows must be considered larger than for the vast majority of flows for which Darcy's law is a valid approximation.

The mathematical properties of the governing partial differential equation for fully saturated continuum ground water flow (Equation [2.5]) are well understood. The form of the equation is typical of a wide variety of physical problems and so has been studied extensively in many contexts. Because the equation is linear, many powerful tools of mathematical analysis are applicable. Exact, analytical solutions are available for a wide variety of problems characterized by very simple geometries, boundary conditions, initial conditions, and parameter fields (usually homogeneous). These analytical solutions are essential in testing and verifying approximate numerical solution techniques, and they often provide considerable insight into more complex problems. In cases where prediction of detailed velocity or concentration fields is unnecessary, analytical solutions often provide adequate precision for certain problems and goals.

For those problems in which the simplifications necessary for attaining analytical solutions are inappropriate (there are many), numerical approximation techniques are highly developed and widely available. A sophisticated literature exists, including several texts devoted exclusively to ground water flow problems (Huyakorn and Pinder, 1983; Remson et al., 1971; Wang and Anderson, 1982). Numerical accuracy and its control are well understood. A number of well-documented, robust, and flexible computer codes are readily available (Bachmat et al., 1980; see also information from the International Ground Water Modeling Center, Indianapolis, Indiana). In addition, in the last few years a variety of computational and graphical tools have been introduced, such as pre- and post-processors and expert systems, designed to aid in the application of such codes. Proper use of numerical codes, however, still requires considerable training and experience, and it is unlikely that solution procedures will ever be fully automated.

Until recently, most numerical methods have focused on efficient and accurate computation of ground water heads for one- and two-dimensional problems. However, in response to the increased availability of affordable computational power, the last few years have seen significant progress in three-dimensional solution techniques. Such techniques are no longer experimental and are beginning to be used in practice (e.g., Ward et al., 1987). In addition, researchers are now focusing attention on the accurate computation of head gradients (velocities), a task much more challenging than accurate computation of heads (Bear and Verrujt, 1987).

The nature of the parameters appearing in the various forms of the fully saturated ground water flow equation is reasonably well understood. Both the hydraulic conductivity and the specific storage are empirical parameters that arise from the simplifications leading to Darcy's law and a workable statement of continuity (see discussion on ground water flow in Chapter 2). While they are not directly measurable, theoretical and experimental studies have clarified how these parameters depend on the properties of the rock and of the fluid when used to predict flow in laboratory columns and boxes (Bear, 1972, pp. 132–136). Less well understood are the natures of these parameters when used to predict average flows over large distances through heterogeneous geologic deposits. Theoretical studies have explored the relationship between large-scale conductivity (and/or transmissivity) and the variability of local conductivity (cf. Dagan, 1986; Gelhar, 1986), but our understanding remains limited.

The state of the art of fully saturated continuum flow modeling is least well developed in the area of hydrologic characterization. The magnitude of flow parameters and their spatial variability currently remain unpredictable a priori. As discussed in detail in Chapter 6, this unpredictability is a major source of uncertainty in ground water flow modeling today.

It is of course impractical to fully characterize an aquifer's permeability distribution via small-scale permeameter testing, since such tests are conducted on disturbed samples of material and are expensive. In addition, simple correlations between more readily measurable geophysical and soil physical parameters have proven elusive (e.g., Lake and Carroll, 1986, pp. 181–221). Several investigators have carried out detailed studies of the spatial structure of permeability and porosity in ground water environments (e.g., Byers and Stephens, 1983; Hoeksema and Kitanidis, 1985; Smith, 1981; Sudicky, 1986). Petroleum engineers have devoted considerable attention to oil and gas reservoir characterization, developing both techniques and insight that are useful to hydrogeologists. These studies have shown that ground water geologic environments are highly variable, but in general, the quantitative knowledge remains very limited.

Parameter values must in general be inferred from field observations of head response to stress. Well tests are the most obvious example. In most applications of ground water flow models, parameter values are obtained via calibration using some type of "inverse technique," leavened by well test estimates and geologic knowledge. Parameter values are chosen that yield satisfactory predictions of observed head at selected observation points (usually few in number) under known conditions.

A large body of theoretical literature has grown up around the ground water "inverse problem" (Yeh, 1986). It is known from these studies that parameter values estimated in this way are nonunique and are very sensitive to errors in measured head data. A number of automated techniques have been suggested for dealing with these problems and for quantifying the resulting uncertainties. There is, however, no agreement on the best approach to this problem, nor is there reason to expect such agreement. Automated techniques remain experimental in practice, and most calibrations proceed by trial-and-error fitting procedures with no quantification of uncertainty. In the hands of an experienced, knowledgeable hydrogeologist, trial-and-error techniques can yield satisfactory results for problems requiring modest spatial resolution. Parameter estimation remains, however, one of the crucial challenges in successful saturated continuum flow modeling, especially for problems of contaminant transport.

Because of the inherent uncertainty in defining the parameter fields for ground water models, a number of investigators have begun exploring the uncertainty in head and velocity predictions that results from parameter uncertainty (e.g., Dagan, 1982; Smith and Freeze, 1979a,b). While limited by a number of restrictive assumptions, these studies are showing how predictive uncertainties may be related to uncertainties in parameters and boundary conditions. They also suggest that prediction uncertainties can be large, especially for velocities. While not yet common in practice, techniques for the quantification of uncertainty are gradually becoming more accepted and accessible.

Implications for the Use of Saturated Continuum Flow Models in Decisionmaking

Because of the relatively long history of development and use of saturated continuum flow models, the issues surrounding their application to modern decisionmaking problems are generally not conceptual or theoretical, but are practical. As noted in the previous section, the physical processes controlling saturated flow are well understood, and the mathematical models describing these processes have been studied extensively. The challenges posed by practical application arise in situations where it is not feasible to model a flow system at the spatial and/or temporal scale appropriate to conceptual and mathematical understanding. Saturated continuum flow models rest on fluid mechanical principles and laboratory column validation of Darcy's law. Field application at this scale is not possible; we lack complete data sets, and if we had such data sets, modeling costs in time and computer resources would be extraordinarily high. Therefore, successful application of ground water flow models rests on the skill and art of the hydrogeologist in understanding when, where, and how to simplify and respond to a lack of information. The next few paragraphs summarize the most important issues that must be addressed in applying saturated continuum flow models to practical ground water problems, given the current state of the art.

Spatial Dimensionality

Many ground water flow problems may be successfully addressed by assuming that flow occurs in only one or two dimensions, i.e., in a single direction or in a plane. Several texts provide careful discussions of the implications of such an assumption (e.g., Bear, 1972; Freeze and Cherry, 1979). The computational savings are obvious. The cost of such a simplification is that model parameters are defined as spatial averages of the "fundamental" parameters (e.g., transmissivity is a depth-average of hydraulic conductivity) and that the predicted responses (head and/or velocity) are similarly averaged. The utility of a reduced dimension model (from a three-dimensional reality) is generally greatest for problems focusing on spatially averaged predictions (volumetric flow rates and/or heads in wells with long screens) and away from boundaries and stresses (wells, for example). Fully three-dimensional flow models typically justify their expense only for problems requiring significant resolution in the vicinity of complex geometries (of boundaries or heterogeneities) or physically small sources or sinks.

Boundary Conditions

The uniqueness of any particular ground water flow problem is expressed in a model in part by locating model boundaries and defining conditions along those boundaries (see Chapter 2). Often the boundaries correspond to physical boundaries in the environment along which conditions are known or can be estimated by the use of data. In many other situations, model boundaries must be defined on the basis of practicality—physical boundaries are unknown or are at great distance from the region of interest. In either case, boundary condition specification is extremely important in many problems and requires a thorough understanding of the mathematical role of boundary conditions as well as the hydrogeologic environment. Boundary condition misspecification is an often overlooked source of significant error (Franke et al., 1987). No matter how complex the model, proper application will always depend on a knowledgeable, trained user working with data that have been collected in such a way as to shed light on boundary conditions.

Transient Versus Steady State

Another valuable assumption in the application of saturated continuum flow models is that of steady state, i.e., that conditions remain constant over time. Because the stresses that drive ground water flow often vary only slowly in time—much more slowly than the system requires to respond—steady state assumptions are often justified. However, there are situations where transients must not be ignored. For example, Sykes et al. (1982) have suggested that ignoring seasonal periodicities in ground water flow direction can lead to otherwise unexpected dispersion during transport. Pulsed-pumping remedial schemes, which are becoming more common, also may demand explicit consideration of transient effects if accurate prediction of contaminant breakthrough is required.

Discretization

The accuracy of numerical approximations to the mathematical equations used to represent ground water flow depends on the size of the discretization used relative to the rate at which the gradient of hydraulic head changes. In other words, if the gradient of head varies rapidly because of hydraulic property heterogeneity or boundary conditions, then discretization must be fine to achieve comparable accuracy to coarse discretization in regions of slowly varying gradient. The choice of grid discretization (in space and/or in time) is further complicated by the averaging incorporated into numerical models. Most models make some very simple assumption about how parameter values vary between computational nodes. For example, many models assume parameters to be constant over a grid block. This averaging, in the face of geologic heterogeneity, requires careful consideration of the resolution required to answer a particular question. Details of flow behavior below the scale of the discretization are usually lost. A general rule of thumb is that the discretization must be at least as dense as the data available for defining parameter heterogeneity. Because data are usually scarce, considerations of numerical accuracy will often determine the grid discretization necessary in many practical problems.

Velocity Computation

More and more problems of saturated continuum flow focus on the prediction of ground water flow velocities. As noted earlier, the accurate prediction of head gradients, on which velocities directly depend, is much more difficult than the accurate prediction of head alone. Computed head gradients are more sensitive to numerical errors of approximation. In addition, head gradients are more sensitive to parameter values. Thus problems requiring accurate velocity prediction, e.g., transport problems, may require more sophisticated numerical methods and may require more careful specification of parameter values for sufficient accuracy.

Parameter Values

The dominant problem in the application of saturated continuum flow models, given today's state of the art, is the specification of parameter values, i.e., the characterization of the geologic environment. Direct measurement is at best costly. Complete characterization is in any case impossible because nondestructive measurement methods are not available. Parameter identification via inversing techniques is computationally difficult. However, when appropriate data are available, the approach provides a practical way to characterize large-scale distributions, for example, hydraulic conductivity distributions from hydraulic head data. While a tremendous amount of current research is focused on these problems, simple panaceas do not seem to be on the horizon. Parameter evaluation will remain a challenging task demanding education, experience, skill, and wisdom.

FLOW IN THE UNSATURATED ZONE

This report is primarily devoted to a discussion of various issues related to modeling water flow and contaminant transport in the saturated zone. A detailed discussion of flow and transport in the unsaturated zone would seem out of place and not necessary. However, the unsaturated zone is the region through which contaminants must pass to reach the saturated zone. The various processes occurring within this region, therefore, play a major role in determining both the quality and the quantity of water recharging into the saturated zone. It is necessary to understand the role played by the unsaturated zone in ground water contamination and how the processes in this zone are either similar to or different from those in deeper flow systems.

Characterization of the Unsaturated Zone

Of course, the major feature of the unsaturated zone that distinguishes it from the saturated zone, as the terms clearly indicate, is the degree of saturation of the pore spaces, in this case by water. In the saturated zone, all of the pores are filled with water (or other water-miscible or immiscible liquids) and the volumetric water content (Θ) is equal to porosity (ϵ). In contrast, the fluid phase occupying the pore spaces in the unsaturated zone may be liquids (mostly water and sometimes nonaqueous-phase liquids [NAPLs]) and gases. The degree of liquid saturation at a given time varies considerably depending on the soil's physical properties (primarily pore-size distribution, which is related to soil texture and structure), and the pattern of inputs and losses of water at the soil surface.

A brief examination of what happens as a completely saturated soil gradually becomes unsaturated is necessary. It is the largest pores that become air-filled first, and removal of water from the smaller pores becomes increasingly more difficult (i.e., requires more work or energy). This phenomenon may be explained by considering the capillary forces that are responsible for water retention in pores.

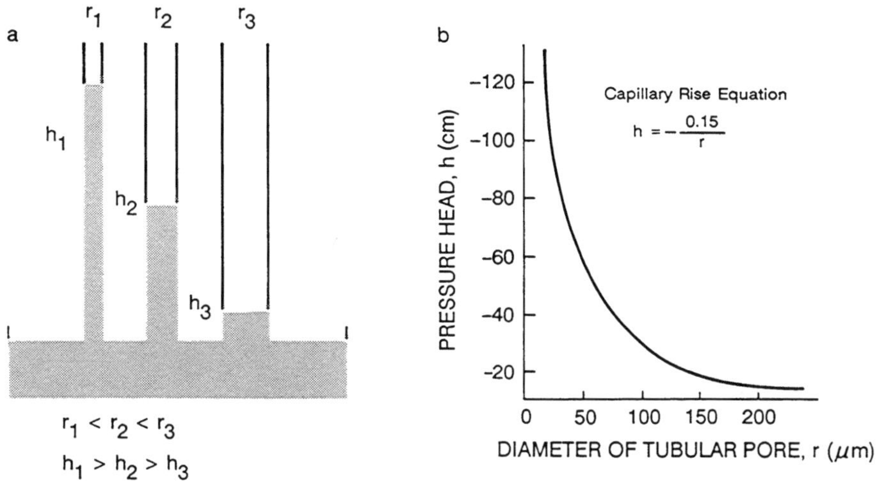

FIGURE 3.1 Relationship between pore size (r) on capillary rise and pressure head (h).

When capillaries of different sizes are placed in water in a beaker, water will rise to different levels above the free water surface. If these capillary tubes are then lifted out of the water, they will not drain unless external pressure is applied. The *capillary* (or suction) forces that hold water inside the capillary against the gravitational forces arise from the attraction of water molecules for each other (cohesion) and the attraction of water molecules to the walls of the capillaries (adhesion).

The height to which water rises in a capillary is indeed a measure of the capillary forces. These forces are stronger in the smaller capillaries, as reflected by the higher rise of water there than in the larger capillaries. In fact, the capillary forces are *inversely* proportional to the radius (r) of the capillary (see Figure 3.1). Water inside a capillary tube (and, by analogy, in soil pores) is under suction and is further illustrated by the concave curvature of the water-air interface at both ends of the capillary tube when it has been taken out of the water. By convention, free water is taken as the reference and is assigned a value of zero for the capillary forces; thus capillary forces are assigned a value less than zero, which is why they are referred to as "suction" forces.

If a porous medium can be thought of as a random network of capillary tubes of varying sizes, it can be seen that the suction force with which water is held in different pore sequences varies inversely

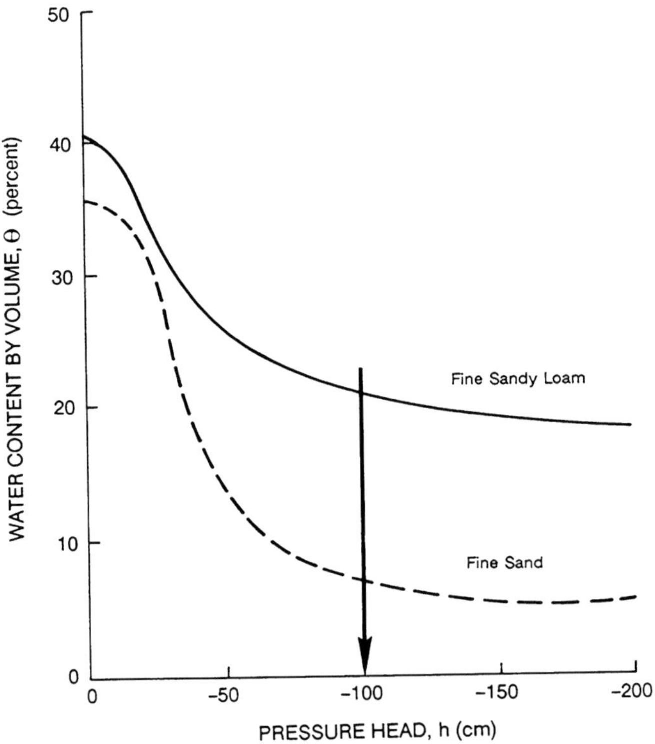

FIGURE 3.2 Examples of soil-water characteristic curves $\Theta(h)$, for several soils. The vertical arrow at h = -100 cm indicates soil water content at field capacity (Θ_{fc}).

with pore radius (r). Thus sandy soils with larger pores retain less water at a given suction than do clayey soils that have smaller pores. This relationship between soil water content (Θ) and negative capillary pressure or suction (h) is an important physical property of soils and is commonly referred to as the soil-water characteristic curve ($\Theta(h)$); typical curves for several soils are shown in Figure 3.2.

Another important physical property of unsaturated soils is their ability to transmit fluids, in particular, water. This property, the hydraulic conductivity (K), has a maximum value, called the saturated hydraulic conductivity (K_s), in a completely saturated soil and decreases dramatically with decreasing soil water content. Again, by using the analogy of capillaries, it is known that for a given hydraulic (or potential) gradient, the flux of water (q) through a capillary is directly proportional to the radius squared (r^2); this principle is known as Poiseuille's law. Thus larger capillaries conduct water much faster than do smaller ones. At saturation, sandy soils will have larger values of K_s than do clayey soils. As a soil becomes unsaturated, the larger pores drain and water flow is restricted to increasingly smaller pore sequences, which conduct water at much lower rates. Based on Poiseuille's law, soil hydraulic conductivity is expected to decrease. This is indeed the case, as shown in Figure 3.3 for the relationship between K and h for several soils. It should be noted that even though sandy soils are much more permeable than clayey soils at saturation, the reverse may be true when these soils are unsaturated.

The knowledge of soil-water characteristic curves ($\Theta[h]$), and soil hydraulic conductivity curves ($K[h]$), is essential for describing water flow and contaminant transport in the unsaturated zone. Because K and Θ are both functions of h, K can be stated as a function of Θ or h. These relationships have been experimentally measured for a large number of soils, and empirical or theoretically based equations for $\Theta(h)$, $K(h)$, and $K(\Theta)$ have been derived to predict them. Some of these are listed in Tables 3.1 and 3.2.

CONCEPTS OF WATER FLOW IN THE UNSATURATED ZONE

As in a saturated soil, the rate and the direction of water flow in an unsaturated soil also depend upon the magnitude of the soil hydraulic conductivity ($K[h]$) and the magnitude and direction of the hydraulic potential gradient (∇H). Darcy's law—which states that soil water flux (q) is directly proportional to potential gradient—is also applicable to unsaturated soils. It should be recognized, however, that the soil hydraulic conductivity now is a strong nonlinear function of h (see Figure 3.3 and Table 3.2) and not a constant value as is the case for saturated flow. In unsaturated soils, the rate of soil water movement may be small even though there may be a large potential gradient, because the hydraulic conductivity ($K[h]$) is small.

Another distinction in water flow between saturated and unsaturated zones must be appreciated. While water flow in the saturated zone usually occurs at a steady rate (except for short periods of time when pumping is initiated), water flow in the unsaturated zone is unsteady (or transient). That is, soil water flux (q) is constant in space (x,y,z) and time (t) for saturated water flow, but may vary

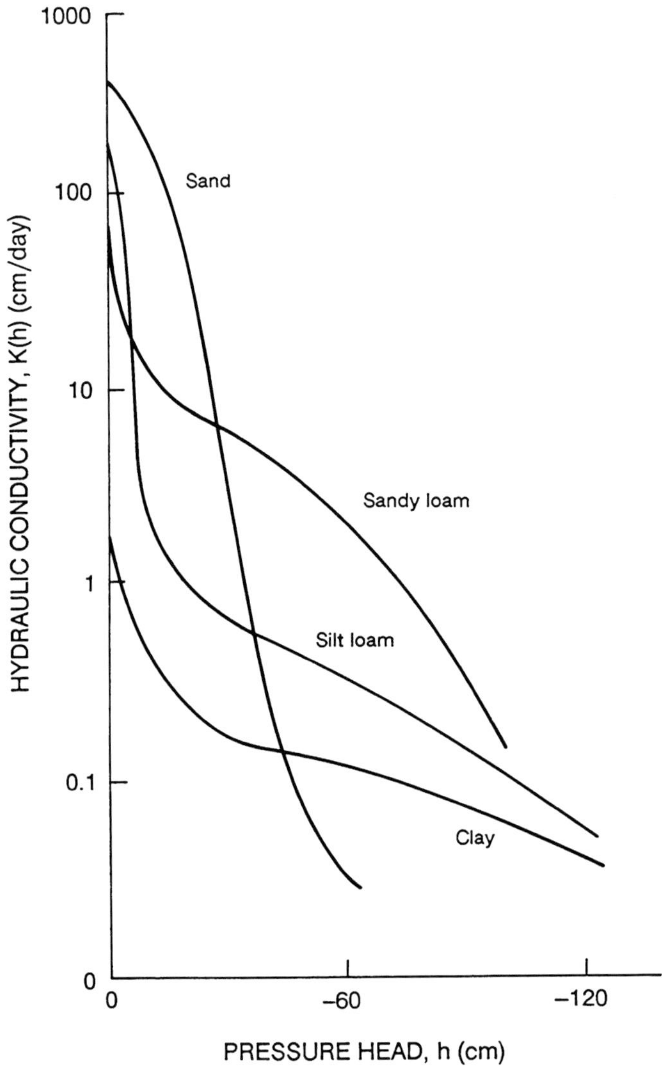

FIGURE 3.3 Examples of $K(h)$ relationships for several soils.

dramatically with both space and time for unsaturated, transient water flow.

A detailed treatment of unsaturated water flow is beyond the scope of the present discussion; the reader is referred to several textbooks on the topic (e.g., Campbell, 1985; Hanks and Ashcroft, 1980; Hillel, 1980a,b; Koorevaar et al., 1983) for a thorough analysis of is

TABLE 3.1 Some $\Theta(h)$ Relationships Reported in the Literature

Function	Source
$\Theta(h) = \Theta_r + \dfrac{[a(\Theta_s - \Theta_r)]}{[a + \lvert h \rvert^b]}$; $h < 0$	Brutsaert (1966)
$\Theta(h) = \Theta_s$; $h \geq 0$	
$\Theta(h) = \Theta_r + \dfrac{[a(\Theta_s - \Theta_r)]}{[a + (\ln \lvert h \rvert)^b]}$; $h < -1$	Haverkamp et al. (1977)
$\Theta(h) = \Theta_s$; $h \geq -1$	
$\Theta(h) = \Theta_r + \dfrac{[\Theta_s - \Theta_r]}{[1 + a(\lvert h \rvert)^b]^m}$; $h < 0$	van Genuchten (1980)
$\Theta(h) = \Theta_s$; $h \geq 0$	
$m = 1 - \dfrac{1}{b}$	
$\Theta(h) = \Theta_r + [\Theta_s - \Theta_r]\,[a/\lvert h \rvert]^b$; $h < a$	
$\Theta(h) = \Theta_s$; $h \geq 0$	

NOTE: Θ_s is saturated water content, equal to porosity; Θ_r is "residual" water content; h is soil-water matric potential; and a, b, and m are constants.

the conceptual basis and theoretical approaches. Here the focus is on a qualitative, phenomenological description of unsaturated water flow. It may be convenient to think of the following three sequential phases in transient water flow: (1) infiltration, (2) redistribution, and (3) static. Phase 1 begins with the input of water at the ground surface (as a result of ponding water, for example), and water begins to infiltrate the soil. The rate of water intake at the ground surface, the infiltration rate (i), is controlled by the method and the rate of water input and the antecedent soil-water conditions. In early stages of phase 1, the infiltration rate is controlled primarily by the water input (application) rate, but in later stages the soil hydraulic properties control the infiltration rate. For infiltration into a "dry" soil, the infiltration rate is initially large because the large hydraulic potential gradients can sustain a large soil water flux. However, as the soil becomes saturated and the wetting front penetrates deeper, the potential gradient driving water flow decreases, and the infiltration rate asymptotically approaches the saturated conductivity (K_s) value. Several equations derived to describe the time dependence of infiltration are listed in Table 3.3.

The above scenario and the equations listed in Table 3.3 are applicable only when water is ponded at the surface. The changes in infiltration rate with time during a water application at a steady rate (as in sprinkler irrigation or during a steady rain) are slightly

TABLE 3.2 Some $K(\Theta)$ or $K(h)$ Relationships Reported in the Literature

Function	Source
$K(h) = a\|h\|^{-b}$	Wind (1955)
$K(h) = K_s[\exp(-a\|h\|)]$	Gardner (1958)
$K(h) = [(\|h\|/a)^b + 1]^{-1}$	Gardner (1958)
$K(h) = [\|h\|/\|h_{cr}\|]^{-b};\ \|h\| \leq \|h_{cr}\|$	Brooks and Corey (1966)
$K(h) = K_s;\ \|h\| \geq \|h_{cr}\|$	
$K(\Theta) = \left[K_s\left(\frac{\Theta - \Theta_r}{\Theta_s - \Theta_r}\right)^b\right];\ b = 3.5$	Averjanov (1950)
$K(h) = \exp\{a[\|h\| - \|h_{cr}\|]\};\ \|h_1\| \leq \|h\| \leq \|h_{cr}\|$	Rijtema (1965)
$K(h) = K_s;\ \|h\| \geq \|h_{cr}\|$	
$K(h) = K_1[\|h\|/\|h_1\|]^{-b};\ \|h\| < \|h_1\|$	
$K(h) = K_s\left[\frac{a}{a + \|h\|^b}\right],\ \begin{matrix} h < 0 \\ h \geq 0 \end{matrix}$	Haverkamp et al. (1977)
$K(\Theta) = a[\exp(b)(\Theta(h))];\ \Theta < \Theta_s$	van Genuchten (1980)
$K(\Theta) = K_s\Theta = \Theta_s$	
$K(h) = K_s[\exp(bh)];\ h < 0$	
$K(h)(K_s);\ h \geq 0$	
$K(h) = a[(\theta(h))^b]$	

NOTE: $K(h)$ is hydraulic conductivity at h; $K(\Theta)$ is hydraulic conductivity at Θ; Θ is volumetric water content; h is hydraulic potential (and has negative values); K_s is saturated hydraulic conductivity; Θ_s is saturated water content; Θ_r is "residual" water content; and a, b, h_{cr}, and h_1 are empirical constants.

TABLE 3.3 Some Expressions Developed to Describe the Time Dependence of Infiltration Rate (i) Following Ponding of Water

Equation	Source
$i = i_c + (b/I)$	Green and Ampt (1911)
$i = Bt^{-n}$	Kostiakov (1932)
$i = i_c + (i_o - i_e)e^{-kt}$	Horton (1940)
$i = i_e + (3/2)t^{-1/2}$	Philip (1957)
$i = i_c + a(M - I)^n;\ I \leq M$	Holton (1961)
$i = i_c;\ I > M$	

NOTE: i is the infiltration rate (cm^3 of water per cm^2 area per hour); I is the cumulative volume (cm^3) of water infiltrated in time t; a, B, M, S, n, and k are constants; i_c is the steady-state infiltration rate; i_e is the initial infiltration rate; and i_o is the final infiltration rate.

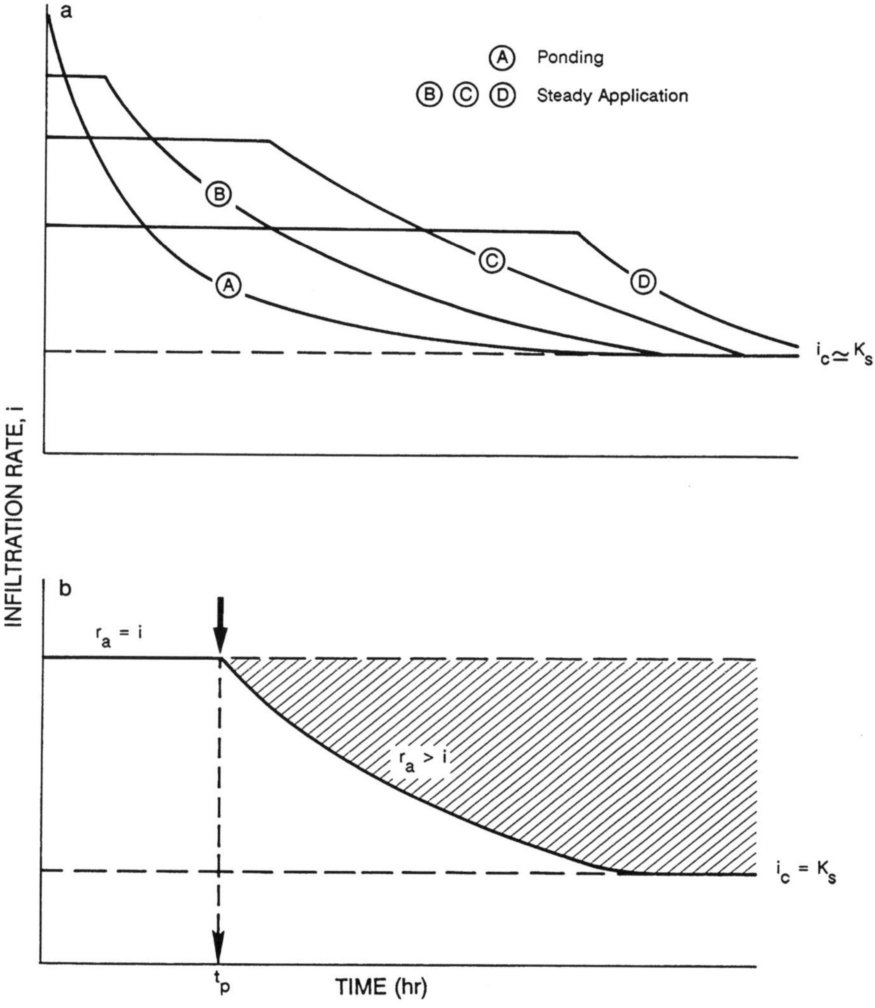

FIGURE 3.4 Examples of the changes in infiltration rate (i) with time for different water application scenarios. Vertical arrows indicate initiation of ponding and surface runoff.

different, although the physical principles that govern flow are the same. The capacity of the soil profile to take in water is initially large enough such that the infiltration rate is equal to the application rate (r_a). If r_a is larger than K_s, soil near the ground surface becomes saturated after some time t_p and water begins to pond at the surface. The infiltration rate begins to drop off and finally reaches the asymptotic value of K_s, as is the case for ponding. In Figure 3.4, several examples illustrating this sequence of events are presented. The physical significance of t_p should be recognized, because it is the time after which the infiltration rate is smaller than the application rate, and the excess water that is ponded on the surface is lost as surface runoff. Note that (1) surface runoff is initiated only when the application rate exceeds the infiltration rate; (2) t_p is small when $r_a \gg K_s$ and gets larger as the value of r_a approaches K_s; and (3) the soil will remain unsaturated ($\Theta < \epsilon$) if $r_a < K_s$, and there will be no surface runoff.

Phase 2 starts when water application at the surface has ceased, and soil water content decreases as water begins to redistribute deeper within the soil profile. During this phase, soil water flux decreases with time to zero in an exponential manner until a unit hydraulic gradient (i.e., only gravitational forces operative) is essentially achieved. This may occur within a few hours in sandy soil but may take several days or even weeks in a slowly permeable clay soil. The soil water content at this time is referred to as the "field-capacity" value in the soil science literature. The presence of a shallow water table, or impermeable soil layers, has a strong impact on the duration of phase 2.

During phase 3, the soil water flux is practically zero (i.e., not measurable) and any decreases in soil water content below the "field-capacity" value are primarily due to losses of water by evaporation and plant uptake (transpiration). Phase 3 continues until the next event of water input (e.g., rainfall, irrigation). In the presence of shallow water table and/or as evapotranspiration proceeds, there may actually be upward water flow, rather than downward flow as assumed for phase 2.

The foregoing scenario, simplified for the present discussion, should illustrate that transient water flow in the unsaturated zone, particularly in the top several meters, is episodic as determined by water input events at the ground surface (see Figure 3.5). Each of these episodes, in turn, has at least three distinct phases where the soil water content and soil water flux vary considerably. Each episode does not have to go through all three successive stages described above; another water input event (e.g., rain, irrigation) may interrupt a given stage. This episodic nature of water flow contrasts with the steady water flow in the saturated zone. At lower depths within the unsaturated zone, beyond the influence of transient conditions at the ground surface, soil water flow is likely to occur under unsaturated

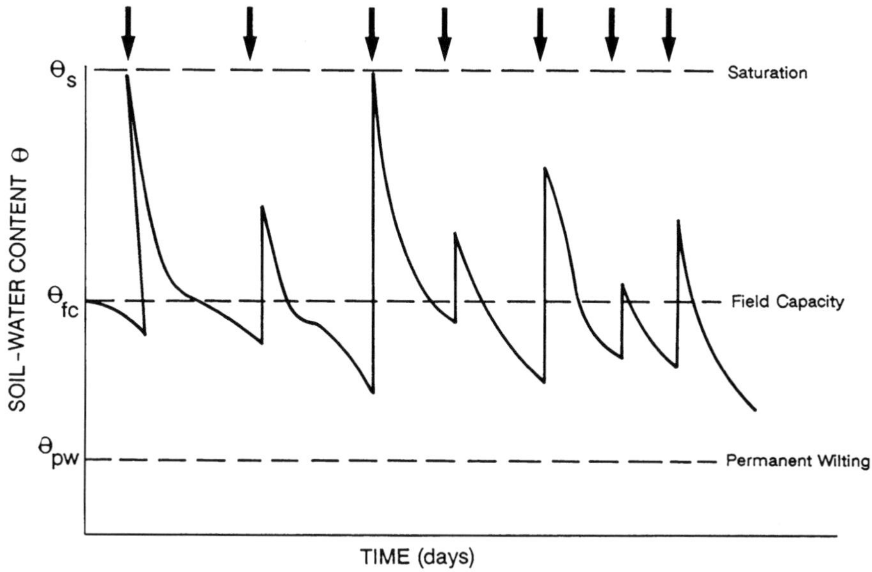

FIGURE 3.5 Variations in soil water content as a result of several episodes of water inputs at the surface. Water inputs are indicated by vertical arrows. Water contents at saturation, field capacity, and permanent wilting are indicated by horizontal dashed lines. Water depletion below Θ_{fc} is the result of losses via evapotranspiration.

conditions, but the soil water flux may be steady and the soil water content constant. The steady flux is equivalent to the annual rate of ground water recharge, which is dependent on site conditions (rainfall patterns, soil physical properties, land use, and so on).

The major differences between saturated and unsaturated flow are summarized in Table 3.4. As noted in Chapter 2, by coupling Darcy's law with the conservation of mass principles, the governing differential equation for transient water flow can be derived; this equation is known as the Richards equation.

FRACTURE FLOW

The term fracture is a general one referring to the various types of discontinuities that can break a medium into blocks (Torsaeter et al., 1987). In the most general case, fracturing adds secondary porosity to some original porosity. The rock blocks contain pores having lengths and widths of about the same dimension and a highly tortuous pattern of interconnection (Shapiro, 1987). The fractures provide more continuous openings with lengths far in excess of their widths. The most widely used conceptual model for fractured media (Figure 3.6) usually considers that discontinuities are represented by joints, fracture zones, and shear zones. As the figure shows, joints are usually discontinuous in their own plane or, in other words, have a definite length and width. When several closely spaced families (sets) of joints are present, they can form a highly interconnected, three-dimensional network for flow (Gale, 1982). Within a given rock or sediment a relatively large number (e.g., five or six) of joint sets can be present, each with its own unique orientation in space. Gale (1982) defines fracture zones as zones of closely spaced and highly interconnected discrete fractures that are generally not filled with clay or other material. The typical width of a fracture zone is from 1 to 10 m. Shear zones provide another example of a large-scale discontinuity with a permeability that can be either higher or lower than the rock mass depending upon filling materials, age, and stress (Gale, 1982).

The picture of fracturing presented so far is one end member in a more complicated hierarchy of multiple-porosity systems. In the case of soluble bedrock like limestone, dolostone, or evaporites, conduit flow can develop as original fracture systems are enlarged by solution. The important feature of conduit flow, when it is able to develop, is the integration of the drainage network (Quinlan and Ewers, 1985). In many ways, the network is analogous to a river system with smaller tributaries supplying water to a succession of

TABLE 3.4 Major Differences Between Saturated and Transient Unsaturated Water Flow in Porous Media

Parameter	Saturated	Unsaturated
Water content (Θ)	$\Theta = \Theta_s$	$0 < \Theta < \Theta_s$
Total potential (H)	$H = h + z\ (h \geq 0)$	$H = h + z\ (-\infty < h < 0)$
Dominant force	Gravity (z)	Matric potential (h)
Driving force	$\partial H/\partial x \simeq$ order (1)	$\partial H/\partial x \simeq$ order (1–10^4)
Hydraulic conductivity (K)	$K = K_s$	$K = K(h)$ or $K(\Theta)$ Range 3 to 8 orders of magnitude
Flux (q)	$q = -K_s \partial H/\partial x$	$q = -K(h)\ \partial H/\partial x$
Governing flow equation	Laplace equation $\nabla^2 h = 0$	Richards equation $\frac{\partial \Theta}{\partial t} = \nabla[K(h)\ \nabla H]$

NOTE: Appreciation is extended to D. Nofziger, Oklahoma State University, Stillwater, for assistance in preparing this table.

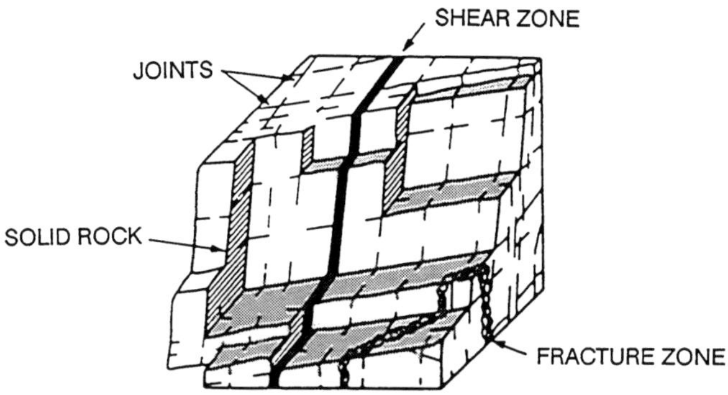

FIGURE 3.6 Conceptualization of discontinuities in a fractured medium. SOURCE: Gale, 1982.

larger and larger conduits. As a result of the integration, both the conduit system and the individual conduits can become large. The famous karst system at Mammoth Cave, Kentucky, is a good example.

The relatively large number of geologic processes that can give rise to fracturing (e.g., tectonism, weathering, glacial stresses, or thermal stresses), coupled with the tendency for older fractures to be propagated into unfractured units, means that fractures are a dominant element controlling fluid migration in all kinds of geologic settings. Similarly, the abundance of carbonate bedrock in North America implies that cases of conduit flow are also common.

Theory of Flow in Fractures

The presence of fractures or conduits in geologic units adds a further complexity to understanding fluid flow. Many of the fundamental principles developed in the previous sections of this chapter need to be extended to deal with fractured media. The first major question of how one conceptualizes flow in a single fracture can be addressed with the help of the so-called parallel plate model (Figure 3.7). A fracture is idealized as a planar opening having a constant thickness or aperture. Flow in the fracture is assumed to obey Darcy's law, or

$$q = K_f \delta H / \delta l, \qquad (3.1),$$

where q is the fracture flux (volume flow per unit time per unit length of cross-sectional fracture area $[2b \times 1]$), K_f is the hydraulic

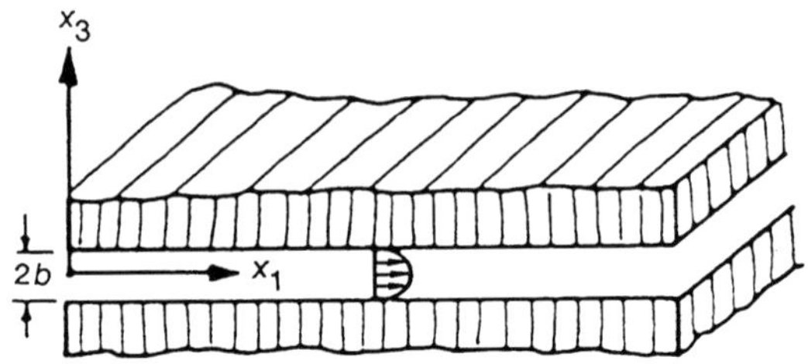

FIGURE 3.7 Idealization of a natural fracture as parallel plates with an aperture of $2b$. SOURCE: Gale, 1982.

conductivity of the fracture, and $\delta H/\delta l$ is the hydraulic gradient along the fracture (Gale, 1982). From fundamental principles of fluid dynamics, K_f can be expressed in terms of fluid and fracture properties as

$$K_f = \frac{\rho g}{12\mu}(2b)^2, \tag{3.2}$$

where ρ is the fluid density, g is gravity, μ is dynamic viscosity, and $2b$ is the fracture aperture. On the basis of (3.2), by keeping in mind that flow occurs through an area $(2b \times 1)$, it can be shown that the quantity of flow passing through a fracture is a function of the aperture cubed (Gale, 1982). This relationship, referred to as the cubic law, is generally valid in describing flow through fractures (Gale et al., 1985). However, indications are that deviations from this general behavior will occur when the fracture surfaces are rough or the fracture surfaces are in contact with one another.

Another feature of fractures that makes them difficult to deal with is the fact that the apertures, and hence the hydraulic conductivity, depend on the stress within the medium. In other words, a fracture can be opened or closed simply by reducing or increasing the forces applied to it. For example, pumping a well in a fractured medium reduces the pore pressure, effectively causing the fracture aperture to decrease. Figure 3.8, taken from Gale (1982), presents experimental data illustrating how fluxes of water through a fracture change with changing stress conditions. Gale (1982) describes a number of empirical-theoretical approaches designed to model the stress coupling to hydraulic conductivity.

In representing the hydraulic conductivity of a fractured medium, it must be considered that two systems of porosity are present. The hydraulic conductivity of the blocks is a straightforward porous medium type—similar to that discussed in Chapter 2. Conceptualizing the hydraulic conductivity of the fractures is more complicated because of the stress dependence.

Recently, there has been interest in formulating and modeling multiphase flow in fractured media. A significant motivation is the assessment of the fractured and unsaturated Yucca Mountain tuff in Nevada as a potential host rock for nuclear waste (Evans and Nicholson, 1987). Further, several of the most serious hazardous waste sites in the United States (e.g., Love Canal and Hyde Park Landfill) involve the migration of NAPLs in fractured media.

As before, the theory for multiphase flow is a generalization of single-phase theory to account for the presence of more than one fluid in the pores and fracture networks. A complication with fractured systems is that relative permeability relationships need to be developed for both the blocks and the fractures. Providing these data is sufficiently difficult that in most cases an attempt is made to approximate these dual-porosity systems as an equivalent single-porosity system.

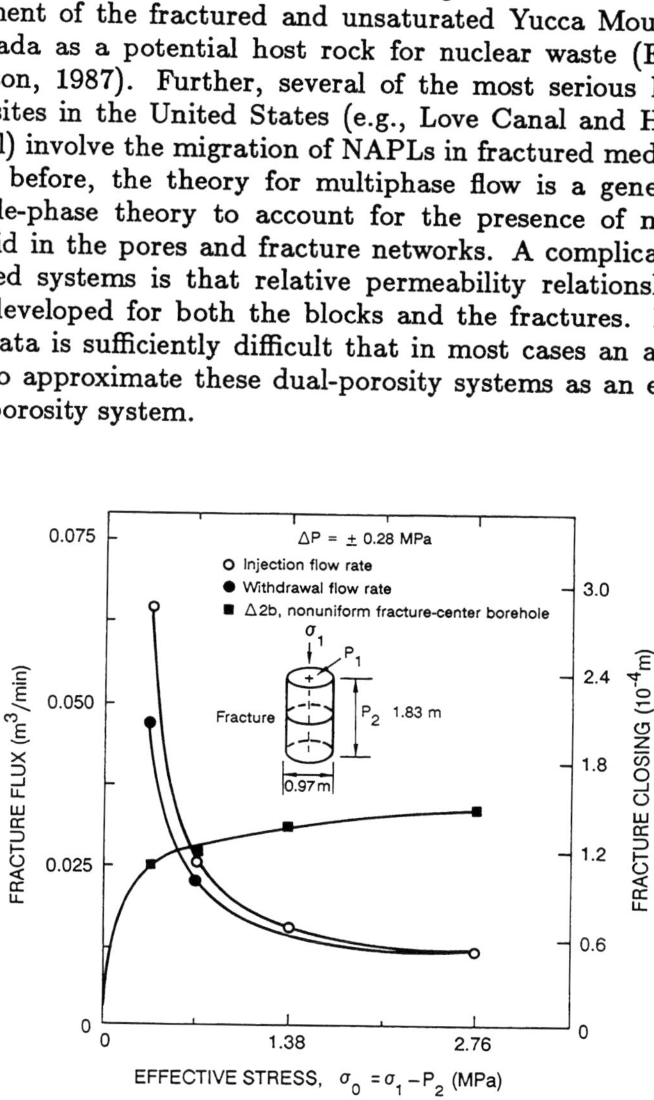

FIGURE 3.8 Experimental data that illustrate how flux in the fracture and aperture change as a function of effective stress for a sandblasted, sawcut fracture surface. SOURCE: Gale, 1982.

So far nothing has been said about the flow of ground water in conduit systems. In general, the character of flow is so very different—often turbulent and partially saturated—that most texts do not treat the basic theory. A few books and papers (LeGrand and Stringfield, 1973; Milanovic, 1981; White, 1969) treat ground water flow in karst and provide a starting point for readers interested in this fascinating topic.

Strategies for Modeling

To date, single-phase or multiphase flow in fractured media has been modeled using one of three possible conceptualizations: (1) an equivalent porous continuum, (2) a discrete fracture network, and (3) a dual-porosity medium. With the first of these approaches, it is assumed that the medium is fractured to the extent that it behaves hydraulically as a porous medium. Under this condition, the continuum equations for porous medium flow developed in Chapter 2 describe the problem mathematically. The actual existence of fractures is reflected in the choice of values for the material coefficients (e.g., hydraulic conductivity, storativity, or relative permeability). Often these parameters take on values significantly different from those used for modeling a porous medium (Shapiro, 1987). Examples of this approach as cited by Shapiro (1987) include Elkins (1953), Elkins and Skov (1960), and Grisak and Cherry (1975).

With the discrete fracture approach, most or all of the ground water moves through a network of fractures. This approach assumes that the geometric character of each fracture (e.g., position in space, length, width, and aperture) is known exactly as well as the pattern of connection among fractures. In the simplest theoretical treatment, the blocks are considered to be impermeable. Figure 3.9a is an idealization of a two-dimensional network of fractures consisting of two different sets. Note how each fracture, represented on the figure by a line segment, has a definite position in space, length, and aperture. The hydraulic characteristics of the fracture system develop as a consequence of the intersection of the individual fractures. In three dimensions, the network can be described in terms of intersecting planes that could be rectangular (Figure 3.9b) or circular in shape (Figure 3.9c). Examples of the discrete fracture treatment of flow in networks include Long et al. (1982, 1985), Robinson (1984), Schwartz et al. (1983), and Smith and Schwartz (1984).

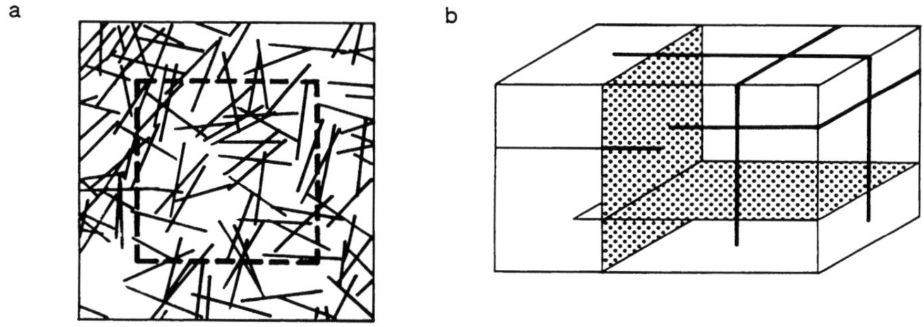

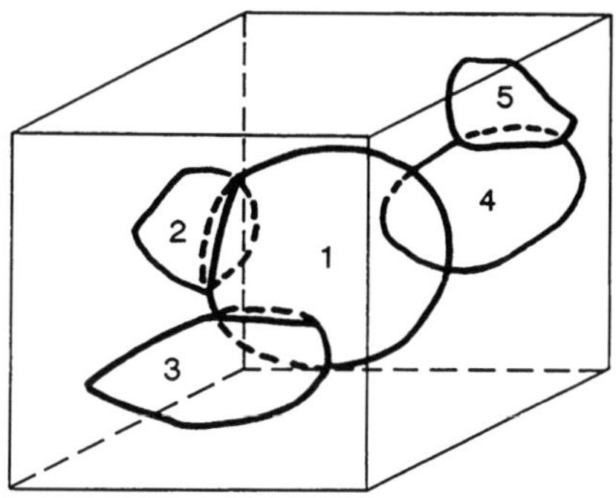

FIGURE 3.9 Three different conceptualizations of fracture networks: (a) a two-dimensional system of line segments (from Shimo and Long, 1987); (b) a three-dimensional system of rectangular fractures (from Smith et al., 1985); and (c) a three-dimensional system of "penny-shaped" cracks (from Long, 1985).

The dual-porosity conceptualization of a fractured medium considers the fluid in the fractures and the fluid in the blocks as separate continua. Unlike the discrete approaches, no account is taken of the specific arrangement of fractures with respect to each other—there is simply a mixing of fluids in interacting continua (Shapiro, 1987). In the most general formulation of the dual-porosity model, the possibility exists for flow through both the blocks and the fractures with a transfer function describing the exchange between the two continua. Mathematically then, one flow equation is written for the fractures and one for the blocks, with the equations coupled by the source-sink terms. Thus a loss in fluid from the fracture represents a gain in fluids in the blocks (Shapiro, 1987).

In most applications involving multiphase flow, a more restrictive approach is followed. Fluids are assumed to flow in the fracture network, with the blocks acting as sources or sinks to the fractures (Torsaeter et al., 1987).

Examining the mathematical details of the dual-porosity formulation is beyond the scope of this overview. To understand the equation development or to locate the available references, readers can refer to Huyakorn and Pinder (1983), Shapiro (1987), and Torsaeter et al. (1987).

ISSUES IN MODELING

The modeling approaches just described are subject to significant complexities of both a theoretical and a practical nature that affect the modeling process. Three main issues are discussed here: (1) whether fractured media can even be approximated as continua, (2) computational constraints on discrete network models, and (3) the uncertainty in establishing the network geometry.

A Fractured Medium as a Continuum

For a porous medium, it is not difficult to believe in the existence of what is termed a representative elemental volume for various controlling parameters. Consider a parameter like hydraulic conductivity as an example. The representative elemental volume is a sample volume for which the hydraulic conductivity is independent of sample volume or averaging volume. In other words, the representative elemental volume exists when a small change in the sample volume does not result in a change in hydraulic conductivity. This concept can be demonstrated using Figure 3.10. When the volume of a porous medium is small (e.g., a few pores), even a slight change in the sample or averaging volume can cause appreciable changes in hydraulic conductivity. As the sample size increases, there comes a point when the hydraulic conductivity is not sensitive to the averaging volume.

In modeling a porous medium as a continuum, it is assumed implicitly that the domain or individual cells within the domain for which the flow equation is written satisfy a representative elemental volume condition. For most cases, this assumption is reasonable. In modeling a fractured medium using continuum approaches, the sample assumption is required. However, in the case of a fractured medium there is much less certainty in the assumption of a representative elemental volume being valid (Schwartz and Smith, 1987).

The first major problem with fractured media is that a representative elemental volume can only be defined when fracture densities are above some critical density. The critical density is defined as that density of fractures that provides connectivity of the network (Figure 3.11). Below the critical density, the network is not connected

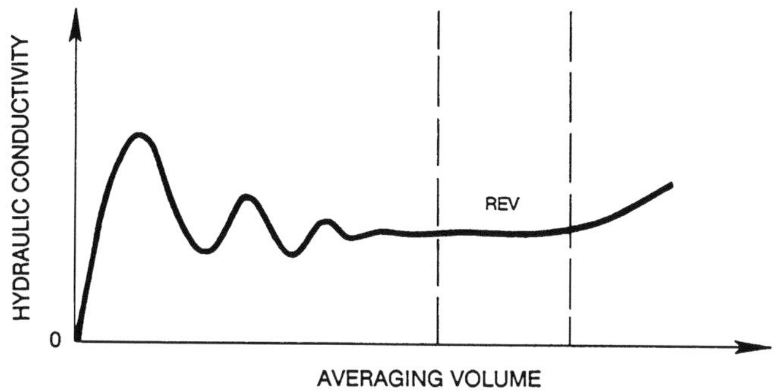

FIGURE 3.10 Variation in hydraulic conductivity as a function of the averaging volume. The dashed lines point to volume where the assumption of a representative elemental volume (REV) is valid. SOURCE: Modified from Shapiro, 1987.

(nonpercolating) and the mean hydraulic conductivity will be zero no matter how large the averaging volume (Schwartz and Smith, 1987). Thus in modeling a fractured system, simply choosing a large volume of rock for a cell will not necessarily guarantee that the assumption of a representative elemental volume is met.

Situations also exist in which the concept of a representative elemental volume is either impractical or invalid. Consider the following two examples. By assuming a network to be connected but sparsely fractured, the averaging volume necessary to obtain a representative elemental volume of the medium could be much larger than the scale of interest. For example, the minimum averaging volume might be a block of rock 200 m on a side, while the scale of interest is 100 m, which makes the concept impractical. Further, a network could be connected, but with a hierarchy of fracture types. In this case, as the sampling volume expands, hydraulic conductivity might keep increasing without necessarily becoming constant.

The concept of a representative elemental volume probably does not hold for some fractured media. Thus there are going to be systems that cannot be modeled by using continuum approaches. Without relatively detailed analyses, it will probably be difficult to identify these systems in advance. When the continuum approaches to modeling are not appropriate, one must turn to discrete modeling approaches, which are clearly not without their own problems, as the next two sections show.

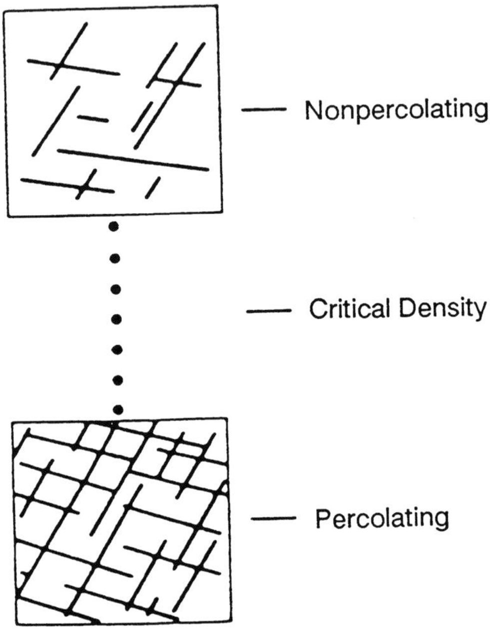

FIGURE 3.11 Examples of percolating and nonpercolating networks in two dimensions. The critical density is the point where infinite clusters of fractures appear and connectivity is achieved. SOURCE: Schwartz and Smith, 1987.

Computational Constraints on Discrete Network Models

Modeling fluid flow in a network of discrete fractures does not require that the fractures behave as a continuum. All of the problem cases discussed in the previous section can be modeled without theoretical constraints. Unfortunately, there are practical constraints that severely limit the capability of modeling discrete fracture networks. The most serious is the number of fracture intersections, because describing flow in the network requires that hydraulic head be calculated at each intersection. In two dimensions, a network with 50,000 intersections will require a major computational effort and will be expensive. Yet the size of a network would in many cases be much smaller than the size of the region of interest. One probably cannot create a discrete fracture system in two dimensions that is large enough to solve intermediate and regional problems. The situation is even more pessimistic when one tries to account for the presence of the rock blocks, or the three-dimensionality of the fracture network. With these more complex flow conditions, one has to further reduce the number of discrete fractures that can be incorporated in the model.

Discrete models are of limited practical value, although they are potentially a theoretically more powerful approach to modeling fractured systems. Their main use to date has been to explore the fundamentals of flow and mass transport in fractured media.

Uncertainty in Establishing the Network Geometry

One further limitation in the use of discrete fracture models is the requirement to specify the exact geometry of the network. There will never be a situation where the geometry of a natural fracture network is exactly known. At best, hydraulic testing can provide estimates of apertures, and fracture mapping in tunnels or on the surface may provide indications of fracture orientation, fracture lengths, and the pattern of connection. However, no tests can provide a definitive description of the network within a rock or sediment mass.

Uncertainty in describing a network ultimately translates into uncertainty in model predictions made for flow in the system. Essentially, the less one knows about a system, the less confident one can be in predicting system response. Stochastic modeling methods (e.g., Smith and Freeze, 1979a) offer a possible approach to making predictions and establishing the potential range of uncertainty in the face of uncertain data.

The complexity of natural fracture networks and the difficulty in making measurements practically guarantee that predictions made by using discrete fracture models will be relatively uncertain. The same situation will probably hold for continuum models as well. However, so little work has been conducted on natural systems that it will require years to fully assess how uncertain predictions in fractured rock systems might be.

Adequacy of Modeling Technology

In most practical problems involving saturated flow in fractured media, there has never been much hesitation in applying continuum-type models. For example, many would argue that the classical methods of well hydraulics appear to model the response of fractured systems to the extent necessary for design. The question remains, however, as to how realistically such models account for the fractured flow processes. Experience from the petroleum industry does suggest that in some cases more sophisticated flow formulations (e.g., dual-porosity models) will be required. Further, issues of stress coupling and the validity of the cubic law will require further study. In addition to the theoretical questions that remain to be resolved, there is a significant gap in practical knowledge about flow in fractured media. For example, only very limited testing has been carried out with well-characterized media, except on the laboratory scale.

The modeling tools exist to deal with fractured media, but at present, results should be interpreted with caution. Systems are often complex and extraordinarily difficult to characterize, especially with the level of effort considered normal for most site investigations. The state of the art in field testing provides a relatively rudimentary estimate of values for some parameters like hydraulic conductivity, while other parameters, like storativity, must be established through fitting simple theoretical models (usually of the porous medium type).

Unsaturated flow modeling of fractured systems is a subject of increasing interest, particularly in light of work at Yucca Mountain in Nevada to assess the feasibility of disposing of high-level nuclear waste in an engineered repository. Most of the same theoretical and practical concerns that were discussed for flow in saturated and fractured media hold for unsaturated media as well. The greatest additional problem lies with the increased difficulty in measuring pertinent parameters down boreholes. According to Evans and Nicholson (1987), a lack of data has restricted the validation of models to a few very simple systems. This problem of the unavailability of data or a large range in variability of existing data was also identified by Pruess and Wang (1987) as an impediment to progress in modeling. Thus the capability in modeling again exceeds the ability to fully establish the validity of the model.

Not much has been written here concerning the multiphase transport of fluids in ground water systems. Notwithstanding the significant capability of solving these problems in petroleum-related applications, there is much less research and overall experience with contaminant-related, multiphase modeling. In the case of dense non-aqueous-phase liquids (DNAPLs), especially the contaminant and petroleum types, problems are sufficiently different that not all of the oil field capabilities are directly transferable. Again, the committee would consider the capability of modeling to exist but without the theoretical and practical experience with the models to consider these applications in any sense routine. As was the case with unsaturated flow modeling in fractured media, limitations in the data provide a further impediment to progress. However, because a variety of organic liquids could be involved, even fewer data are available.

REFERENCES

Averjanov, S. F. 1950. About permeability of subsurface sites in case of incomplete saturation. Engineering Collection, Vol. VII, as quoted by P. Ya, Polubarinova Kuchina in The Theory of Groundwater Movement, English translation by J. M. Roger de Wiest, 1962. Princeton University Press, Princeton, N.J.

Bachmat, Y., J. Bredehoeft, B. Andrews, D. Holtz, and S. Sebastian. 1980. Groundwater Management: The Use of Numerical Models, Water Resources Monograph 5. American Geophysical Union, Washington, D.C., 127 pp.

Bear, J. 1972. Dynamics of Fluids in Porous Media (1988 edition). Dover, New York, 764 pp.

Bear, J., and A. Verrujt. 1987. Modeling Groundwater Flow and Pollution. Reidel, Dordrecht, 414 pp.

Brooks, R. H., and A. T. Corey. 1966. Properties of porous media affecting fluid flow. Journal of the Irrigation Drainage Division of the American Society of Civil Engineers 92(IR2), 61–68.

Brutsaert, W. 1966. Probability laws for pore-size distributions. Soil Science 101, 85–92.

Byers, E., and D. B. Stephens. 1983. Statistical and stochastic analyses of hydraulic conductivity and particle-size in a fluvial sand. Soil Science Society of America Journal 47, 1072–1081.

Campbell, G. S. 1985. Soil Physics with BASIC. Developments in Soil Science No. 14. Elsevier, Amsterdam.

Dagan, G. 1982. Stochastic modeling of groundwater flow by unconditional and conditional probabilities, 1. Conditional simulation and the direct problem. Water Resources Research 18(4), 813–833.

Dagan, G. 1986. Statistical theory of groundwater flow and transport pore to laboratory, laboratory to formation, and formation to regional scale. Water Resources Research 22(9), 120S–134S.

Elkins, L. F. 1953. Reservoir performance and well spacing, Spraberry Trend area field of West Texas. Transactions, American Institute of Mining Engineers 198, 177–196.

Elkins, L. F., and A. M. Skov. 1960. Determination of fracture orientation from pressure interference. Transactions, American Institute of Mining Engineers 219, 301–304.

Evans, D. D., and T. J. Nicholson. 1987. Flow and transport through unsaturated fractured rock: An overview. Pp. 1-10 in Flow and Transport Through Unsaturated Fractured Rock, D. D. Evans and T. J. Nicholson, eds. Geophysical Monograph 42, American Geophysical Union, Washington, D.C.

Franke, O. L., T. E. Reilly, and G. D. Bennett. 1987. Definition of Boundary and Initial Conditions in the Analysis of Saturated Ground-Water Flow Systems—An Introduction, Book 3, Chapter B5, Techniques of Water-Resources Investigations of the United States Geological Survey. U.S. Geological Survey, Reston, Va., 15 pp.

Freeze, R. A., and J. A. Cherry. 1979. Groundwater. Prentice-Hall, Englewood Cliffs, N.J. 604 pp.

Gale, J. E. 1982. Assessing the permeability characteristics of fractured rock. Pp. 163–181 in Recent Trends in Hydrogeology, T. N. Narasimhan, ed. Geological Society of America, Special Paper 189.

Gale, J. E., A. Rouleau, and L. C. Atkinson. 1985. Hydraulic properties of fractures. Congress International Association of Hydrogeologists 17, 1–11.

Gardner, W. R. 1958. Some steady-state solutions to the unsaturated flow equation with application to evaporation from a water table. Soil Science 85, 228–232.

Gelhar, L. W. 1986. Stochastic subsurface hydrology from theory to applications. Water Resources Research 22(9), 135S–145S.

Green, W. H., and G. A. Ampt. 1911. Studies on soil physics, I. Flow of air and water through soils. Journal of Agricultural Science 4, 1–24.

Grisak, G. E., and J. A. Cherry. 1975. Hydrologic characteristics and response of fractured till and clay confining a shallow aquifer. Canadian Geotechnical Journal 12, 23–43.

Hanks, R. J., and G. L. Ashcroft. 1980. Applied Soil Physics. Advanced Series in Agricultural Science, No. 8. Springer-Verlag, New York.

Haverkamp, R., M. Vauclin, J. Touma, P. J. Wierenga, and G. Vachaud. 1977. A comparison of numerical simulation models for one-dimensional infiltration. Soil Science Society of America Journal 41, 285–294.

Hillel, D. 1980a. Fundamentals of Soil Physics. Academic Press, New York.

Hillel, D. 1980b. Applications of Soil Physics. Academic Press, New York.

Hoeksema, J. R., and P. K. Kitanidis. 1985. Analysis of the spatial structure of properties of selected aquifers. Water Resources Research 21(4), 562–572.

Holton, H. N. 1961. A concept of infiltration estimates in watershed engineering. Agricultural Research Service Publication No. 41-51, U.S. Department of Agriculture, Washington, D.C.

Horton. 1940. An approach toward a physical interpretation of infiltration-capacity. Soil Science Society of America, Proceedings 5, 399–417.

Huyakorn, P. S., and G. F. Pinder. 1983. Computational Methods in Subsurface Flow. Academic Press, New York, 473 pp.

Koorevaar, P., G. Menelik, and C. Dirksen. 1983. Elements of Soil Physics. Developments in Soil Science No. 13. Elsevier, Amsterdam.

Kostiakov, A. N. 1932. On the dynamics of the coefficient of water percolation in soils and on the necessity of studying it from a dynamic point of view for purposes of amelioration. Trans. Com. International Soil Science, 6th, Moscow, Russia, Part A, 17–21.

Lake, L. W., and H. B. Carroll, Jr., eds. 1986. Reservoir Characterization. Academic Press, Orlando, Fla., 659 pp.

LeGrand, H. E., and V. T. Stringfield. 1973. Karst hydrology—A review. Journal of Hydrology 19, 1–23.

Long, J. C. S. 1985. Verification and characterization of fractured rock at AECL Underground Research Laboratory. BMI/OCRD-17, Office of Crystalline Repository Development, Battelle Memorial Institute, 239 pp.

Long, J. C. S., J. S. Remer, C. R. Wilson, and P. A. Witherspoon. 1982. Porous media equivalents for networks of discontinuous fractures. Water Resources Research 18, 645–658.

Long, J. C. S., P. Gilmour, and P. A. Witherspoon. 1985. A model for steady fluid flow in random three-dimensional networks of disc-shaped fractures. Water Resources Research 21, 1105–1115.

Milanovic, P. 1981. Karst Hydrogeology. Water Resources Publications, Littleton, Colo., 444 pp.

Philip, J. R. 1957. The theory of infiltration, 4. Sorptivity and algebraic infiltration equations. Soil Science 84, 257–264.

Pruess, K., and J. S. Y. Wang. 1987. Numerical modeling of isothermal and nonisothermal flow in unsaturated fractured rock—A review. Pp. 11–21 in Flow and Transport Through Unsaturated Fractured Rock, D. D. Evans and T. J. Nicholson, eds. Geophysical Monograph 42, American Geophysical Union, Washington, D.C.

Remson, I., G. M. Hornberger, and F. J. Molz. 1971. Numerical Methods in Subsurface Hydrology. Wiley-Interscience, New York, 389 pp.

Rijtema, P. E. 1965. An analysis of actual evapotranspiration. Agriculture Research Report No. 659. Center for Agricultural Publication and Documentation, Wageningen, The Netherlands.

Robinson, P. C. 1984. Connectivity flow and transport in network models of fractured media. TP 1072, Theoretical Physics Division, AERE, Harwell, U.K.

Quinlan, J.F., and R.O. Ewers. 1985. Ground water flow in limestone terrains: Strategy, rationale, and procedures for reliable efficient monitoring of ground water quality in karst areas. Pp. 197–234 in Proceedings of the National Symposium and Exposition on Aquifer Restoration and Ground Water Monitoring. National Water Well Association, Columbus, Ohio.

Schwartz, F. W., and L. Smith. 1987. An overview of the stochastic modeling of dispersion in fractured media. Pp. 729–750 in Advances in Transport Phenomena in Porous Media, J. Bear and M. Y. Corapcioglu, eds. NATO Advanced Study Institutes Series, Series E Applied Sciences, Vol. 128. Martinus Nijhoff Publishers, Dordrecht.

Schwartz, F. W., L. Smith, and A. S. Crowe. 1983. A stochastic analysis of macroscopic dispersion in fractured media. Water Resources Research 19, 1253–1265.

Shapiro, A. M. 1987. Transport equations for fractured porous media. Pp. 407–471 in Advances in Transport Phenomena in Porous Media, J. Bear and M. Y. Corapcioglu, eds. NATO Advanced Study Institutes Series, Series E Applied Sciences, Vol. 128. Martinus Nijhoff Publishers, Dordrecht.

Shimo, M., and J. C. S. Long. 1987. A numerical study of transport parameters in fracture networks. Pp. 121–131 in Flow and Transport Through Unsaturated Fractured Rock, D. D. Evans and T. J. Nicholson, eds. Geophysical Monograph 42, American Geophysical Union, Washington, D.C.

Smith, L. 1981. Spatial variability of flow parameters in a stratified sand. Journal of Mathematical Geology 13, 1–21.

Smith, L., and R. A. Freeze. 1979a. Stochastic analysis of steady state groundwater flow in a bounded domain, 1. One-dimensional simulations. Water Resources Research 15(3), 521–528.

Smith, L., and R. A. Freeze. 1979b. Stochastic analysis of steady state groundwater flow in a bounded domain, 2. Two-dimensional simulation. Water Resources Research 15(6), 1543–1559.

Smith, L., and F. W. Schwartz. 1984. An analysis of fracture geometry on mass transport in fractured media. Water Resources Research 20, 1241–1252.

Smith, L., and R. A. Freeze. 1979a. Stochastic analysis of steady state groundwater flow in a bounded domain, 1. One-dimensional simulations. Water Resources Research 15(3), 521–528.

Smith, L., and R. A. Freeze. 1979b. Stochastic analysis of steady state groundwater flow in a bounded domain, 2. Two-dimensional simulation. Water Resources Research 15(6), 1543–1559.

Smith, L., and F. W. Schwartz. 1984. An analysis of fracture geometry on mass transport in fractured media. Water Resources Research 20, 1241–1252.

Smith, L., C. W. Mase, and F. W. Schwartz. 1985. A stochastic model for transport in networks of planar fractures. Pp. 523–536 in Greco 35 Hydrogeologic. Ministere de la Recherche et la Technologie Centre Nationale de la Recherche Scientifique, Paris.

Sudicky, E. A. 1986. A natural gradient experiment on solute transport in a sand aquifer; spatial variability of hydraulic conductivity and its role in the dispersion process. Water Resources Research 22(13), 2069–2082.

Sykes, J. F., S. B. Pahwa, R. B. Lantz, and D. S. Ward. 1982. Numerical simulation of flow and contaminant migration at an extensively monitored landfill. Water Resources Research 18(6), 1687–1704.

Torsaeter, O., J. Kleppe, and T. van Golf-Racht. 1987. Multiphase flow in fractured reservoirs. Pp. 553–629 in Advances in Transport Phenomena in Porous Media, J. Bear and M. Y. Corapcioglu, eds. NATO Advanced Study Institutes Series, Series E Applied Sciences, Vol. 128. Martinus Nijhoff Publishers, Dordrecht.

van Genuchten, M. T. 1980. A closed-form equation for predicting the hydraulic conductivity of unsaturated soils. Soil Science Society of America Journal 44, 892–898.

Wang, H. F., and M. P. Anderson. 1982. Introduction to Groundwater Modeling. Freeman, San Francisco, 237 pp.

Ward, D. S., D. R. Buss, J. W. Mercer, and S. S. Hughes. 1987. Evaluation of a groundwater corrective action at the Chem-Dyne hazardous waste site using a telescopic mesh refinement modeling approach. Water Resources Research 23(4), 603–617.

White, F. M. 1974. Viscous Fluid Flow. McGraw-Hill, New York, 725 pp.

White, W. B. 1969. Conceptual models for carbonate aquifers. Ground Water 7(3), 15–22.

Wind, G. P. 1955. Flow of water through plant roots. Netherlands Journal of Agricultural Science 3, 259–264.

Yeh, W. W-G. 1986. Review of parameter identification procedures in groundwater hydrology: The inverse problem. Water Resources Research 22(2), 95–108.

4

Transport

INTRODUCTION

Ground water contamination occurs when chemicals are detected where they are not expected and not desired. This situation is a result of movement of chemicals in the subsurface from some source (perhaps unknown) that may be located some distance away. Ground water contamination problems are typically advection dominated (see "Dissolved Contaminant Transport" in Chapter 2), and the primary concerns in defining and treating ground water contamination problems must initially focus on physical transport processes. If a contaminant is chemically or biologically reactive, then its migration tends to be attenuated in relation to the movement of a nonreactive chemical. The considerations of reaction add another order of magnitude of complexity to the analysis of a contamination problem, in terms of both understanding and modeling. Regardless of the reactivity of a chemical, a basic key to understanding and predicting its movement lies in an accurate definition of the rates and direction of ground water flow.

The purpose of a model that simulates solute transport in ground water is to compute the concentration of a dissolved chemical species in an aquifer at any specified place and time. Numerical solute transport models were first developed about 20 years ago. However, the modeling technology did not have a long time to evolve before a great demand arose for its application to practical and complex field problems. Therefore the state of the science has advanced from theory to practice in such a short time (considering the relatively small number of scientists working on this problem at that time) that a large base of experience and hypothesis testing has not accumulated. It appears that some practitioners have assumed that the underlying theory and numerical methods are further beyond the research, development, and testing stage than they actually are.

Most transport models include reaction terms that are mathematically simple, such as decay or retardation factors. However, these do not necessarily represent the true complexities of many reactions. In reality, reaction processes may be neither linear nor equilibrium controlled. Rubin (1983) discusses and classifies the chemical nature of reactions and their relation to the mathematical problem formulation.

Difficult numerical problems arise when reaction processes are highly nonlinear, or when the concentration of the solute of interest is strongly dependent on the concentration of numerous other chemical constituents. However, for field problems in which reactions significantly affect solute concentrations, simulation accuracy may be limited less by mathematical constraints than by data constraints. That is, the types and rates of reactions for the specific solutes and minerals in the particular ground water system of interest are rarely known and require an extensive amount of data to assess accurately. Mineralogic variability may be very significant and may affect the rate of reactions, and yet be essentially unknown. There are very few documented cases for which deterministic solute transport models have been applied successfully to ground water contamination problems involving complex chemical reactions.

Many contaminants of concern, particularly organic chemicals, are either immiscible or partly miscible with water. In such cases, processes in addition to those affecting a dissolved chemical may significantly affect the fate and movement of the contaminant, and the conventional solute transport equation may not be applicable. Rather, a multiphase modeling approach may be required to represent phase composition, interphase mass transfer, and capillary forces, among other factors (see Pinder and Abriola, 1986). This would concurrently impose more severe data requirements to describe additional parameters, nonlinear processes, and more complex geochemical and biological reactions. Faust (1985) states, "Unfortunately, data such as relative permeabilities and capillary pressures for the types of fluids and porous materials present in hazardous waste sites are not readily available." Well-documented and efficient multiphase models applicable to contamination of ground water by immiscible and partly miscible organic chemicals are not yet generally available.

TRANSPORT OF CONSERVATIVE SOLUTES

Much of the recently published research literature on solute transport has focused on the nature of dispersion phenomena in ground water systems and whether the conventional solute transport equation accurately and adequately represents the process causing changes in concentration in an aquifer. In discussing the development and derivation of the solute transport equation, Bear (1979, p. 232) states, "As a working hypothesis, we shall assume that the dispersive flux can be expressed as a Fickian type law." The dispersion process is thereby represented as one in which the concentration gradient is the driving force for the dispersive flux. This is a practical engineering approximation for the dispersion process that proves adequate for some field problems. But, because it incorrectly represents the actual physical processes causing observed dispersion at the scale of many field problems, which is commonly called macrodispersion, it is inadequate for many other situations.

The dispersion coefficient is considered to be a function both of the intrinsic properties of the aquifer (such as heterogeneity in hydraulic conductivity and porosity) and of the fluid flow (as represented by the velocity). Scheidegger (1961) showed that the dispersivity of a homogeneous, isotropic porous medium can be defined by two constants. These are the longitudinal dispersivity and the transverse dispersivity of the medium. Most applications of transport models to ground water contamination problems documented to date have been based on this conventional formulation, even when the porous medium is considered to be anisotropic with respect to flow.

The consideration of solute transport in a porous medium that is anisotropic would require the estimation of more than two dispersivity parameters. For example, in a transversely isotropic medium, as might occur in a horizontally layered sedimentary sequence, the dispersion coefficient would have to be characterized on the basis of six constants. In practice, it is rare that field values for even the two constants longitudinal and transverse dispersivity can be defined uniquely. It appears to be impractical to measure as many as six constants in the field. If just single values of longitudinal and transverse dispersivity are used in predicting solute transport in an anisotropic medium when the flow direction is not always parallel to the principal directions of anisotropy, then dispersive fluxes will be either overestimated or underestimated for various parts of the flow system. This can sometimes lead to significant errors in predicted concentrations.

Dispersion and advection are actually interrelated and are dependent on the scale of measurement and observation and on the scale of the model. Because dispersion is related to the variance of velocity, neglecting or ignoring the true velocity distribution must be compensated for in a model by a correspondingly higher value of dispersivity. Domenico and Robbins (1984) demonstrate that a scaling up of dispersivity will occur whenever an $(n-1)$ dimensional model is calibrated or used to describe an n-dimensional system. Davis (1986) used numerical experiments to show that variations in hydraulic conductivity can cause an apparently large dispersion to occur even when relatively small values of dispersivity are assumed. Similarly, Goode and Konikow (1988) show that representing a transient flow field by a mean steady-state flow field, as is commonly done, inherently ignores some of the variability in velocity and must also be compensated for by increased values of dispersivity.

The scale dependence of dispersivity coefficients (macrodispersion) is recognized as a limitation in the application of conventional solute transport models to field problems. Anderson (1984) and Gelhar (1986) show that most reported values of longitudinal dispersivity fall in a range between 0.01 and 1.0 on the scale of the measurement (see Figure 4.1). Smith and Schwartz (1980) conclude that macrodispersion results from large-scale spatial variations in hydraulic conductivity and that the use of relatively large values of dispersivity with uniform hydraulic conductivity fields is an inappropriate basis for describing transport in geologic systems. It must be recognized that geologic systems, by their very nature, are complex, three-dimensional, heterogeneous, and often anisotropic. The greater the degree to which a model approximates the true heterogeneity as being uniform or homogeneous, the more must the true variability in velocity be incorporated into larger dispersion coefficients. We will never have so much hydrogeologic data available that we can uniquely define all the variability in the hydraulic properties of a geologic system; therefore, assumptions and approximations are

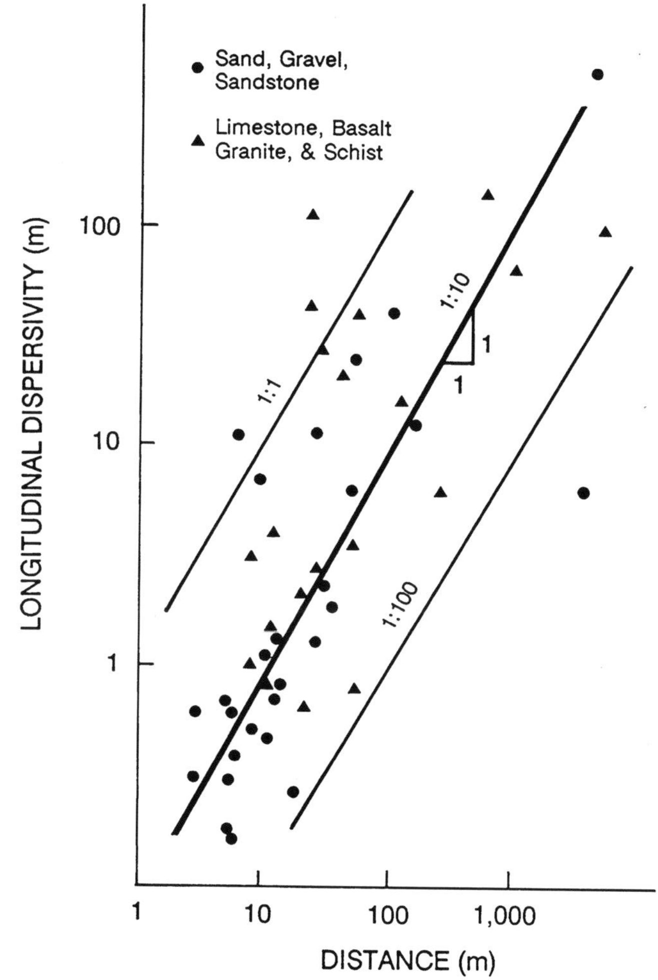

FIGURE 4.1 Variation of dispersivity with distance (or scale of measurement). SOURCE: Modified from Anderson, 1984.

always necessary. Clearly, the more accurately and precisely we can define spatial and temporal variations in velocity, the lower will be the apparent magnitude of dispersivity. The role of heterogeneities is not easy to quantify, and much research is in progress on this problem.

An extreme but common example of heterogeneity is rocks that exhibit a dominant secondary permeability, such as fractures or solution openings. In these types of materials, the secondary permeability channels may be orders of magnitude more transmissive than the porous matrix of the bulk of the rock unit. In these settings, the most difficult problems are identifying where the fractures or solution openings are located, how they are interconnected, and what their hydraulic properties are. These factors must be known in order to predict flow, and the flow must be calculated or identified in order to predict transport. Anderson (1984) indicates that where transport occurs through fractured rocks, diffusion of contaminants from fractures to the porous rock matrix can serve as a significant retardation mechanism, as illustrated in Figure 4.2. Modeling of flow and transport through fractured rocks is an area of active research, but not an area where practical and reliable approaches are readily available. Modeling the transport of contaminants in a secondary permeability terrain is like predicting the path of a hurricane without any knowledge of where land masses and oceans are located or which way the earth is rotating.

Because there is not yet a consensus on how to describe, account for, or predict scale-dependent dispersion, it is important that any conventional solute transport model be applied to only one scale of a problem. That is, a single model, based on a single value of dispersivity, should not be used to predict both near-field (near the solute source) and far-field responses. For example, if the dispersivity value that is used in the model is representative of transport over distances on the order of hundreds of feet, it likely will not accurately predict dispersive transport on smaller scales of tens of feet or over

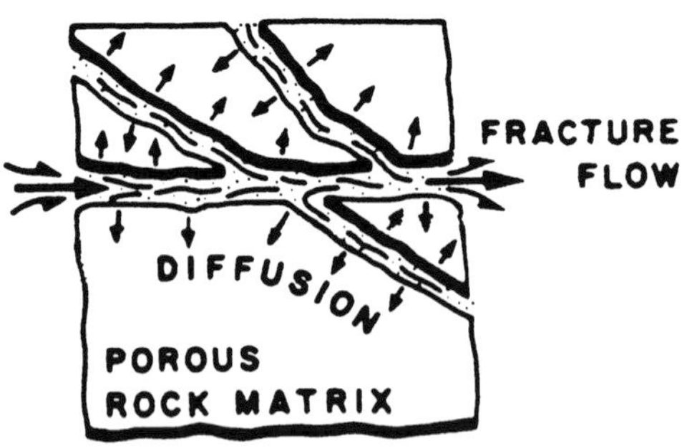

FIGURE 4.2 Flow through fractures and diffusion of contaminants from fractures into the rock matrix of a dual-porosity medium. SOURCE: Anderson, 1984.

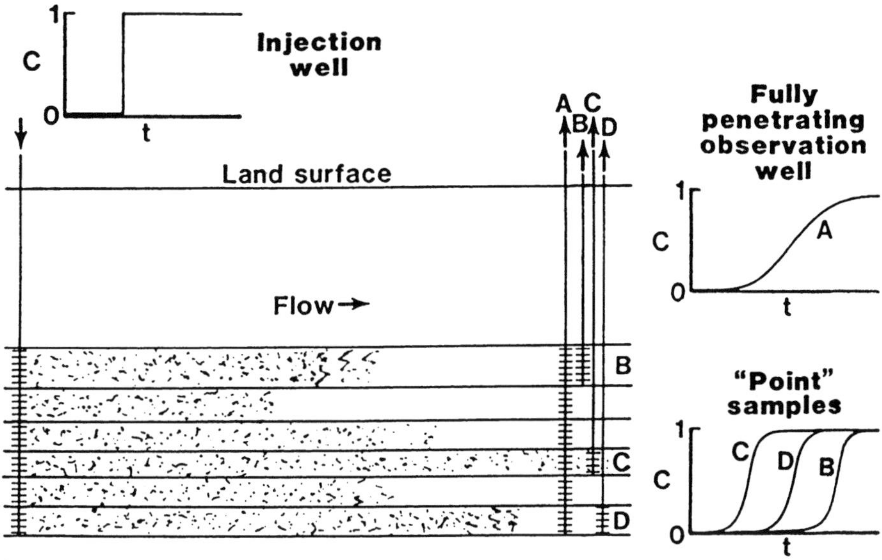

FIGURE 4.3 Effect of sampling scale on estimation of dispersivity. SOURCE: L. F. Konikow, U.S. Geological Survey, Reston, Va., written communication, 1989.

larger scales of miles. Warning flags must be raised if measurements of parameters such as dispersivity are made or are representative of some scale that is different from that required by the model or by the solution to the problem of interest.

Similarly, the sampling scale and manner of sampling or measuring dependent variables, such as solute concentration, may affect the interpretation of the data and the estimated values of physical parameters. For example, Figure 4.3 illustrates a case in which a tracer or contaminant is injected into a confined and stratified aquifer system. It is assumed that the properties are uniform within each layer but that the properties of each layer differ significantly. Hence, for injection into a fully penetrating injection well, as shown at the left of Figure 4.3, the velocity will differ between the different layers. Arrival times will then vary at the sampling location. Samples collected from a fully penetrating observation well will yield a gentle breakthrough curve indicating a relatively high dispersivity. However, breakthrough curves from point samples will be relatively steep, indicating low dispersivity in each layer. The finer scale of sampling yields a more accurate conceptual model of what is really happening, and an analogous model should yield more reliable predictions.

Because advective transport and hydrodynamic dispersion depend on the velocity of ground water flow, the mathematical simulation model must solve at least two simultaneous partial differential equations. One is the flow equation, from which velocities are calculated, and the other is the solute transport equation, which describes the chemical concentration in ground water. If the range in concentration throughout the system is small enough that the density and viscosity of the water do not change significantly, then the two equations can be decoupled (or solved separately). Otherwise, the flow equation must be formulated and solved in terms of intrinsic permeability and fluid pressure rather than hydraulic conductivity and head, and iteration between the solutions to the flow and transport equations may be needed.

Ground water transport equations, in general, are more difficult to solve numerically than are the ground water flow equations, largely because the mathematical properties of the transport equation vary depending upon which terms in the equations are dominant in a particular situation (Konikow and Mercer, 1988). The transport equation has been characterized as "schizophrenic" in nature (Pinder and Shapiro, 1979). If the problem is advection dominated, as it is in most cases of ground water contamination, then the governing partial differential equation becomes more hyperbolic in nature (similar to equations describing the propagation of a shock front or wave propagation). If ground water velocities are relatively low, then changes in concentration for that particular problem may result primarily from diffusion and dispersion processes. In such a case, the governing partial differential equation is more parabolic in nature. Standard finite-difference and finite-element methods work best with parabolic and elliptic partial differential equations (such as the transient and steady-state ground water flow equations). Other approaches (including method of characteristics, random walk, and related particle-tracking methods) are best for solving hyperbolic equations. Therefore no one numerical method or simulation model will be ideal for the entire spectrum of ground water contamination problems encountered in the field. Model users must take care to use the model most appropriate to their problem.

Further compounding this difficulty is the fact that the ground water flow velocity within a given multidimensional flow field will normally vary greatly, from near zero in low-permeability zones or near stagnation points, to several feet per day in high-permeability areas or near recharge or discharge points. Therefore, even for a single ground water system, the mathematical characteristics of the transport process may vary between hyperbolic and parabolic, so that no one numerical method may be optimal over the entire domain of a single problem.

A comprehensive review of solute transport modeling is presented by Naymik (1987). The model survey of van der Heijde et al. (1985) reviews a total of 84 numerical mass transport models. Currently, there is much research on mixed or adaptive methods that aim to minimize numerical errors and combine the best features of alternative standard numerical approaches because none of the standard numerical methods is ideal over a wide range of transport problems.

In the development of a deterministic ground water transport model for a specific area and purpose, an appropriate level of complexity (or, rather, simplicity) must be selected (Konikow, 1988). Finer resolution in a model should yield greater accuracy. However, there also exists the practical constraint that even when appropriate data are available, a finely subdivided three-dimensional numerical transport model may be too large or too expensive to run on available computers. This may also be true if the model incorporates nonlinear processes related to reactions or multiphase transport. The selection of the appropriate model and the appropriate level of complexity will remain subjective and dependent on the judgment and experience of the analysts, the objectives of the study, and the level of prior information on the system of interest.

In general, it is more difficult to calibrate a solute transport model of an aquifer than it is to calibrate a ground water flow model. Fewer parameters need to be defined to compute the head distribution with a flow model than are required to compute concentration changes using a solute transport model. A model of ground water flow is often calibrated before a solute transport model is developed because the ground water seepage velocity is determined by the head distribution and because advective transport is a function of the seepage velocity. In fact, in a field environment, perhaps the single most important key to understanding a solute transport problem is the development of an accurate definition (or model) of the flow system. This is particularly relevant to transport in fractured rocks where simulation is based on porous-media concepts. Although the head distribution can often be reproduced satisfactorily, the required velocity field may be greatly in error.

It is often feasible to use a ground water flow model alone to analyze directions of flow and transport, as well as travel times, because contaminant transport in ground water is so strongly (if not predominantly) dependent on ground water flow. An illustrative example is the analysis of the Love Canal area, Niagara Falls, New York, described by Mercer et al. (1983). Faced with inadequate and uncertain data to describe the system, Monte Carlo simulation and uncertainty analysis were used to estimate a range of travel times (and the associated probabilities) from the contaminant source area to the Niagara River. Similarly, it is possible and often useful to couple a particle-tracking routine to a flow model to represent advective forces in an aquifer and to demonstrate explicitly the travel paths and travel times of representative parcels of ground water. This ignores the effects of dispersion and reactions but may nevertheless lead to an improved understanding of the spreading of contaminants.

Figure 4.4 illustrates in a general manner the role of models in providing input to the analysis of ground water contamination problems. The value of the modeling approach lies in its capability to integrate site-specific data with equations describing the relevant processes as a basis for predicting changes or responses in ground water quality. There is a major difference between evaluating existing contaminated sites and evaluating new or planned sites. For the former, if the contaminant source can be reasonably well defined, the history of contamination itself can, in effect, serve as a surrogate long-term tracer test that provides critical information on velocity and dispersion at a regional scale. However, it is more common that when a contamination problem is recognized and investigated, the locations, timing, and strengths of the contaminant sources are for the most part unknown, because the release to the ground water system occurred in the past when there may have been no monitoring. In such cases it is often desirable to use a model to determine the characteristics of the source on the basis of the present distribution of contaminants. That is, the requirement is to run the model backward in time to assess where the contaminants came from. Although this is theoretically possible, in practice there is usually so much uncertainty in the definition of the properties and boundaries of the ground water system that an unknown source cannot be uniquely identified. At new or planned sites, historical data are commonly not available to provide a basis for model calibration and to serve as a control

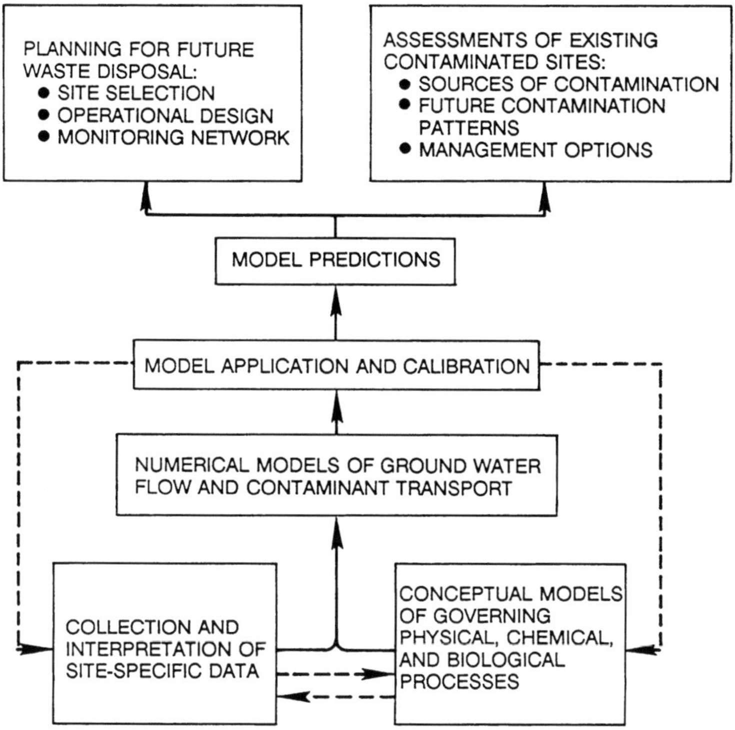

FIGURE 4.4 Overview of the role of simulation models in evaluating ground water contamination problems. SOURCE: Konikow, 1981.

on the accuracy of predictions. As indicated in Figure 4.4, there should be allowances for feedback from the stage of interpreting model output both to the data collection and analysis phase and to the conceptualization and mathematical definition of the relevant governing processes.

NONCONSERVATIVE SOLUTES

The following sections assess the state of the art for modeling abiotic transformations, transfers between phases, and biological processes in the subsurface. Descriptions of all of these processes are provided in Chapter 2. The focus for this assessment is an examination of what reactions are important and to what extent they can be described by equilibrium and kinetic models.

Equilibrium and Kinetic Models of Reactions

Much of the discussion in this section refers to inorganic species. Generally speaking, reactions can be described from an equilibrium and/or kinetic viewpoint. As an example of an equilibrium description, consider the following reversible reaction:

$$A + B = 2C. \tag{4.1}$$

At equilibrium, the reaction is described by the following mass law:

$$K = \frac{(C)^2}{(A)(B)}|\text{at equilibrium}, \tag{4.2}$$

where

- K = the equilibrium constant, which is temperature dependent
- A,B = reactant species
- C = product species
- ( ) = activity, a thermodynamic property that is proportional to the aqueous-phase concentration for dissolved species and to the partial pressure for a gas

This equation implies that at equilibrium the activities of the reactants and products should be related in the relative proportions indicated by (4.2). When this relationship is not achieved, mass is transferred forward in the reverse reaction.

Another way of looking at a reaction involves using a kinetic approach. Unlike the equilibrium description, the kinetic approach describes how the concentration of a constituent changes with time. Kinetics usually are expressed by a rate law of the form

$$r_A = -kV[A]^x[B]^y, \tag{4.3}$$

where

- r_A = rate of mass accumulation of species A ($M_A T^{-1}$)
- V = volume of water in the system being modeled (L^3)
- k = a rate coefficient that depends on the mechanisms of the reaction, the temperature, and other environmental conditions and has units giving $M_A T^{-1}$ for r_A
- x,y = reaction-order exponents

Expressions like r_A can be employed directly as source-sink terms in the mass balance equations used in a solute transport model.

Whether one adopts an equilibrium or a kinetic model clearly depends on the character of the reaction. For example, irreversible reactions cannot be described using the equilibrium concept because they continue until all of the reactions are used up. Another factor determining how a reaction can be discussed is the rate of the reaction relative to the physical transport process. For example, when a reversible reaction is fast in relation to advection and dispersion, an equilibrium description is appropriate. When the reaction is slower, a kinetic viewpoint is more appropriate. Thus the same reaction can be described in different ways depending upon the conditions of transport.

Abiotic Reactions

Table 4.1 summarizes the equilibrium relationships for each of the abiotic transformations and the status of the thermodynamic databases describing these reactions. Except for radioactive decay, which is an irreversible reaction and is not describable using equilibrium concepts, a simple mass law expression describes the reaction. Further, the databases of thermodynamic parameters (i.e., E^o, K_a, K_{so}, K_{sb}, and K_{sa}) are relatively complete and accessible only for oxidation/reduction and acid/base processes. Therefore fundamental knowledge must be generated to extend the thermodynamic databases.

Table 4.2 provides an assessment of the kinetic relationships for the abiotic transformations. Two mechanisms, radioactive decay and acid/base processes, are well understood and have well-defined databases. The decay coefficients for all the major radionuclides have been known for some time. For acid/base processes, the reactions normally are so fast that instantaneous equilibrium can be assumed. The term for r_{ab} in Table 4.2 reflects that the formation of A^- ion is proportional to the rate of change in concentration of the sum of acidic (HA) and basic (A^-) species multiplied by the fraction of the total composed by A^-.

For three mechanisms—dissolution, complexation, and substitution/hydrolysis—the means to describe the rate of reaction have been identified, but the kinetic parameters needed to quantify the rates (i.e., A, k_d, k_{com}, k_H, K_{OH}, and K_N) are poorly defined for many of the relevant species and conditions. Fundamental information

TABLE 4.1 Summary and Evaluation of Thermodynamics of Abiotic Transformation Mechanisms for Aqueous Species

Mechanisms/Reaction Form	Thermodynamic Relation	Status Code[a]
Radioactive decay, $P \rightarrow D$ + nuclear particle	Not applicable	1
Oxidation/reduction, $N^+ + R \rightarrow N + R^+$	$E = E^o + RT \ln \left[\frac{\{R^+\}\{N\}}{\{R\}\{N^+\}}\right]$	1
Acid/base processes, $HA = H^+ + A^-$	$\frac{\{H^+\}\{A^-\}}{\{HA\}} = K_a$	1
Precipitation/dissolution, $C^+ + A^- = CA(\text{solid})$	$\{C^+\}\{A^-\} = K_{so}$	2
Complexation, $C^+ + L^- = CL$	$\frac{\{CL\}}{\{C\}\{L^-\}} = K_{st}$	2
Substitution/hydrolysis, $RX + N = RN + X$	$\frac{\{RN\}\{X\}}{\{RX\}\{N\}} = K_{su}$	2

[a] "1" indicates that the thermodynamic database is well established, and "2" that the database is incomplete.

NOTE: Definition of parameters: P = parent radionuclide, D = daughter product of decay, R = reductant or electrophile, N^+ = oxidant or nucleophile, R^+ = oxidized reductant, N = reduced oxidant, HA = acid, H^+ = hydrogen ion, A^- = conjugant base of HA or anion, C^+ = cation, CA(solid) = precipitate, L^- = ligand, CL = complex, X = leaving group, E = potential (volts), E^o = standard potential (volts), K_a = acid/base dissociation constant, K_{so} = solubility product, K_{st} = stability constant, K_{su} = substitution constant, R = universal gas constant = 1.99×10^{-3} kcal/mole K, T = temperature (in kelvin units).

needed to implement the kinetic relationships in models awaits future research.

The kinetic formulations for two mechanisms, oxidation/reduction and precipitation, are not firmly established. The equations presented in Table 4.2 are reasonable first approaches, but considerably more research is needed before the correct forms are known for these two reactions. It is likely that the correct forms will not be universally generalizable and that the correct form will vary depending on the species involved and on environmental conditions. It also is clear that many oxidation/reduction reactions are irreversible and are at

TABLE 4.2 Summary and Evaluation of Kinetics of Abiotic Transformation Mechanisms

Mechanism	Kinetic Expression	Status Code[a]
Radioactive decay	$r_{rd} = -\lambda[P]V$	1
Oxidation/reduction	$r_{red} = -k_{red}[N^+][R]V$	3
Acid/base processes	Instantaneous equilibrium $r_{ab} = \left(\frac{K_a}{[HA] + [A^-]}\right)\left(\frac{d([HA] + [A^-])}{dt}\right)V$	1
Precipitation	$r_p = -k_pA(1 - Q/K_{so})[C^+]^n$	3
Dissolution	$r_d = k_dA(1 - Q/K_{so})^n$	2
Complexation	$r_{com} = k_{com}[C^+][L^-]V$	2
Substitution/hydrolysis	$r_{sub} = -k_T[RX]V$, where $k_T = k_H[H^+] + k_{OH}[OH^-] + k_N$	2

[a] "1" indicates that the kinetic expression is well understood and the database of kinetic parameters is well established; "2" indicates that the kinetic expression is well understood and the database of kinetic parameters is incomplete; "3" indicates that the kinetic expression is poorly understood and the kinetic parameters are incomplete.

NOTE: Definition of parameters: r_{rd} = rate of loss of parent isotope by radioactive decay (MT^{-1}), λ = decay constant (T^{-1}), r_{red} = rate of loss of reactants (N^+ and R) due to oxidation/reduction (MT^{-1}), k_{red} = oxidation/reduction rate coefficient ($L^3M^{-1}T^{-1}$), r_{ab} = rate of formation of A^- ion due to acid/base reaction (MT^{-1}), r_p = rate of loss of C^+ due to precipitation (MT^{-1}), k_p = rate coefficient for precipitation (units depend on n), n = exponent ≥ 1, A = surface area onto which solid forms (L^2), r_d = rate of formation of C^+ due to dissolution (MT^{-1}), k_d = dissolution rate coefficient ($ML^{-1}T^{-1}$), r_{com} = rate of formation of complex, k_{com} = complexation rate coefficient ($L^3M^{-1}T^{-1}$), r_{sub} = rate of loss of original electrophile (MT^{-1}), k_T = total substitution rate coefficient (J^{-1}), k_H = acid-catalyzed rate coefficient ($L^3M^{-1}T^{-1}$), k_{OH} = base-catalyzed rate coefficient ($L^3M^{-1}T^{-1}$), and k_N = neutral rate coefficient (T^{-1}).

disequilibrium in ground waters at low temperatures (Lindberg and Runnells, 1984).

Geochemical Models

In reality, many different abiotic transformations occur simultaneously. Equations describing all the different reactions and all the participating chemical species must be solved together because one chemical species can participate in several different reactions of the same type or of different types. Several geochemical models, such as MINTEQ (Felmy et al., 1984), PHREEQE (Parkhurst et al., 1980), GEOCHEM (Sposito and Mattigod, 1980), and WATEQF (Plummer et al., 1976), are designed to set up and solve simultaneous thermodynamic equations for many different reactions and species. These models were first used with purely inorganic chemical systems, but they also are being applied to systems with organic chemicals.

The geochemical models begin with a thermodynamic database for the normally dominant aqueous species present at the normal pH range of waters. A computer code then poses and solves a mass balance problem, subject to the thermodynamic constraint that an equilibrium be reached for all reactions. The equations describing the thermodynamic system are posed in a matrix format in which the stoichiometric coefficients of the chemical reactions form the elements of the matrix. The use of activity coefficients, computed by the model, allows the thermodynamic and mass balance equations to be solved together. The solution proceeds by successive approximations in a stepwise manner from the measured concentrations. Successive iterations continue until all equilibrium expressions are true (e.g., until $Q = K$ for all reactions) and all elemental mass balances sum to the original concentrations within acceptable tolerances. The solution describes the equilibrium makeup of the water if all known reactions proceed to equilibrium.

Most geochemistry models do not model reactions kinetically. They assume that equilibrium models apply, and therefore none of the information contained in Table 4.2 is contained in most geochemistry models. A notable exception is the model Code EQ6, which incorporates some kinetic expressions for dissolution and precipitation of minerals (Delany et al., 1986; Wolery et al., 1988). Because many of the reactions (particularly oxidation/reduction, precipitation, dissolution, substitution/hydrolysis, and some complexation reactions) are slow and often cannot be modeled with instantaneous equilibrium, the output of a geochemistry model provides information only on possible trends. The actual transformations that occur and the times and distances over which they occur are not specified by geochemical models; research in this area is badly needed for predictive modeling.

Incorporation of Abiotic Transformations into Solute Transport Models

Geochemical codes can be used independently of transport codes to provide estimates of species mobility. The incorporation of only one abiotic reaction into a solute transport model normally does not pose extraordinary difficulties, as long as the kinetic expression and its parameters can be specified. Incorporation requires that mass balances be written for all species of interest; in most cases, the species of interest include all reactants and any products of special interest. The mass balance must then be solved, subject to the flow conditions provided externally or from a coupled flow model. Except for very simple cases, such as radioactive decay in a homogeneous aquifer, solution involves a numerical technique. The most difficult aspect usually is keeping mass balances on all reacting species and maintaining electrical charge balance in the solution, while concentrations change over space and time.

For the more complicated and often more realistic situation in which many transformation reactions are possible, the logical step is to link a geochemical model with a mass balance model. The models CHEMTRN (Miller and Benson, 1983) and TRANQL (Cederberg et al., 1985) have achieved operational linkage between chemical equilibrium calculations and modeling of transport through porous media. However, and this is very important, the geochemical models are very complex and computationally demanding to solve for only the equilibrium relationships and relatively simple kinetic expressions. The linking of geochemical models, in their present forms, into solute transport models is difficult because of the computing demand. Therefore most of the currently available geochemistry models may be inappropriate for solute transport modeling, even if kinetics can be included. Instead, simpler versions—perhaps involving only the species and reactions of known importance to the site or to specific problems being studied—need to be developed. This is the approach that has been taken in the new FASTCHEM™ model (Hostetler et al., 1988), in which a minimal chemical database is used in the MINTEQ portion (Felmy et al., 1984; Peterson et al., 1987) of a package of computer codes that include linkage between equilibrium geochemical modeling and hydrologic flow and transport. Another possible approach would be to provide a general framework for equilibrium computations (e.g., some form of simultaneous-equation solver) that could be coupled to appropriate kinetic expressions to provide a flexible source-sink subroutine for chemical reactions.

Before geochemical models can be routinely incorporated into solute transport models, the models must be tested in controlled field studies. So far, researchers have validated only sections or portions of the geochemical codes against field or laboratory data, such as the state of saturation of the water with respect to calcium carbonate. The lack of extensive validation is not the fault of the codes; instead, it points out the surprising lack of laboratory and field studies that are designed for, or are suitable for, testing of theoretical models. Modelers tend to go their own way, building impressive computer codes, while experimentalists tend to gather data for purposes other than evaluating models. The resolution of this problem must eventually come from a close interaction among modelers, experimentalists, and field scientists. We have probably reached the point at which it is now imperative to gather laboratory and field data to evaluate the validity and utility of the geochemical codes.

One of the unique aspects of solute transport modeling for subsurface waters is the very large amount of solid surface area to which the water is exposed. Surfaces often behave as reactants in the types of reactions described above. In particular, functional groups on solids can act as oxidants or reductants, acids or bases, complexing ligands, and dissolution or precipitation sites. In general, the thermodynamics and kinetics for surface reactions are similar to those for reactions in solution. However, transport of dissolved species to and from the surface often needs to be taken into account; thus the kinetics often are controlled by diffusion processes. In addition, surface reactions are unique because the surface reactants and products need not move with the water phase, but often remain fixed on the solid phase.

Phase Transfers

The transfer of chemical species between two different phases can be a major source-sink term. The major transfers are between the following pairs of phases: solid/liquid, liquid/liquid, liquid/gas, and solid/gas. The simplest way of modeling phase transfers is as equilibrium processes (Table 4.3), in which the concentration or density in one phase is proportional to the concentration or density in the other phase.

The exception to this general rule is the transfer of colloids, which normally is not described in terms of equilibrium. While the forms for the partitioning expressions are well established, only for gas/liquid partitioning is there a relatively complete set of partition coefficients. For the other transfers, the partition parameters are not complete, because they depend on site-specific characteristics (e.g., K_{sorp}, K_{ex}, Q_m, and b) or have not been systematically studied yet. Therefore database expansion and the means to characterize local

TABLE 4.3 Summary and Evaluation of the Thermodynamics of Phase Transfers

Phases Involved	Partitioning Expressions	Status Code[a]
Solid/liquid		
Organic solutes	$Q = K_d[C]^N$	2
Inorganic ions	$K_{ex} = \dfrac{\{C^+{}_1\}(C_2 - X)}{\{C^+{}_2\}(C_1 - X)}$	2
Colloids	Not applicable	—
Liquid/liquid	$[C]_o = K_*[C]$	2
Liquid/gas	$[C]_g = H[C]$	1
Solid/gas	$Q = \dfrac{Q_m b[C]_g}{1 + b[C]_g}$	2

[a]"1" indicates that partitioning parameters are well established and "2" that partitioning parameters are incomplete.

NOTE: Definitions of parameters: Q = sorption density of the solute on and/or in the solid phase (MM_{sol}^{-1}), K_d = sorption constant (units depend on N), N = sorption exponent for Freundlich isotherms, $[C]$ = concentration of solute in liquid phase (usually water) (ML^{-3}), K_{ex} = ion-exchange coefficient, $\{C^+{}_1\}$ and $\{C^+{}_2\}$ = solution activities of two exchanging cations, $(C_1 - X)$ and $(C_2 - X)$ = surface densities of two exchanging cations (ML_x^{-2} or MM_x^{-1}), $[C]_o$ = concentration of species in second (usually organic) liquid phase (ML_o^{-3}), K_* = phase distribution coefficient ($L^3L_o^{-3}$), $[C]_g$ = concentration of volatile species in gas phase (ML_g^{3}), H = Henry's constant ($L^3L_g^{-3}$), Q_m = monolayer sorption density to a solid (ML_{sol}^{-2} or MM_{sol}^{-1}), b = Langmuir energy constant ($L_g^3M^{-1}$).

sites are key needs for the successful modeling of subsurface phase transfers.

Assessment of Kinetics

As noted previously, phase transfers are usually modeled using equilibrium reaction models. Kinetic expressions containing a k_m parameter multiplied by some sort of concentration difference normally are used to describe filtration of colloids but, when necessary, can be used to describe most of the other transfers as well. Kinetic models do require additional information about the system of interest, namely, estimates or measurements of k_m, the interfacial surface area, and concentrations or densities in both phases. These parameters often are difficult to estimate. It is for this reason that the equilibrium models are used to model phase transfers, but even so the parameters needed to quantify them—K_d, K_{ex}, K_*, Q_m, and b—are difficult to estimate for field conditions.

Most mass transport expressions, not including colloid filtration and gas/liquid transfer, are only first approximations. Considerable research on transfer mechanisms will be necessary before reliable expressions are available. Such research is necessary because the instantaneous equilibrium approaches are not appropriate when solute advection is significant, which occurs in highly porous media and near wells and trenches used for remediation.

Incorporation of Phase Transfers into Solute Transport Models

The incorporation of phase transfers into solute transport models is relatively straightforward, as long as the rate term is available. Modeling with instantaneous equilibria is especially easy, because the movement of the solute can be modeled as simple advection at a velocity that is a fraction of the liquid flow velocity. For example, a sorbing organic solute moves at velocity v′:

$$v' = v/(1 + \rho_{aq} K_d/\epsilon, \qquad (4.4)$$

where

- v = water flow velocity (LT^{-1})
- v' = velocity of movement of the center of mass of the solute (LT^{-1})
- ρ_{aq} = mass of aquifer solids per unit volume of aquifer ($M_{sol}L^{-3}$)
- ϵ = porosity, or volume of liquid per unit volume of aquifer (L^3L^{-3})
- K_d = linear partition coefficient ($L^3M_{sol}{}^{-1}$)
- $(1 + \rho_{aq}K_d/\epsilon)$ = retardation factor

Similar relationships can be derived for the other transfers.

Ion exchange is a special case of instantaneous equilibrium, because two competing ions must be modeled in the liquid and solid phases. The task of following two aqueous species and two solid phase species increases the computational burden but has been achieved successfully.

When a mass transport approach is necessary, the computations become more cumbersome because concentrations or densities in both phases must be modeled. However, as long as mass balances on species in both phases are set up and linked via the transport rates, the mass transport approach creates no special modeling difficulties.

The main difficulty with implementing any of the approaches to solute transport with phase transfers is characterization of the subsurface medium in terms of partition coefficients, gas flows, and nonaqueous liquid contents. Because the solid, gas, and second liquid phases do not move with the water, a solute being transported in the water can encounter many different environments for phase transfers. This spatial heterogeneity of nonaqueous phases in subsurface porous media can play an important role in determining the fate of chemical species that transfer across phase boundaries. Gathering the data to characterize the heterogeneous subsurface in terms of its nonaqueous phases is expensive and difficult. The difficulty is compounded when the nonaqueous phase is changing or moving over time, as could be the case when a gas is generated or is in multiphase liquid flow.

Biological Reactions

On the one hand, modeling biological reactions involves all the same considerations and approaches described for abiotic reactions. This similarity occurs because microorganisms are catalysts for the transformations described under abiotic transformations. Microorganisms are especially associated with the oxidation/reduction and substitution/hydrolysis reactions, but they also can catalyze acid/base, precipitation/dissolution, and complexation reactions. Microorganisms can catalyze chemical reactions, but they cannot cause reactions that are not thermodynamically possible; thus microorganisms affect only the rate of reactions.

On the other hand, modeling biological reactions involves features that are not part of abiotic transformations. The two most critical features are (1) that the microorganisms have to grow through the utilization of required substrates and (2) that most of the microorganisms are attached to the solid-medium particles. The first feature means that, in modeling of the biological processes, mass balances often are needed for required substrates, even when they are not the chemical species of interest. The second feature means that the reactions by the microorganisms and the mass balance for the microorganisms must be posed in terms of a solid phase that does not move with the water.

Microbiological Kinetics

For any biodegradable compound in the water, its removal from the water is described by a flux from the liquid and to the microorganisms attached to the solids:

$$r_{\text{bio}} = -JA, \tag{4.5}$$

where

r_{bio} = rate of substrate loss from the pore liquid by biological transformation (MT^{-1})
J = substrate flux to the attached microorganisms ($ML^{-2}T^{-1}$)
A = surface area of microbial biofilm or microcolonies (L^2)

The flux (J) can be computed by simultaneous solution of two equations: one for mass transport of substrate to the surface of the biofilm or microcolony and the other for simultaneous diffusion and utilization of the substrate within the film or colony. Relatively simple techniques are available for obtaining J (Rittmann and McCarty, 1981), as long as the accumulation of attached biomass is known and the transformation kinetics can be characterized, both of which are difficult to determine in a field situation.

The most common kinetic expression for microbial utilization by individual cells is the Monod relation:

$$r_{\text{ut}} = -\frac{kX_a S_f V}{K_s + S_f}, \tag{4.6}$$

where

r_{ut} = rate of substrate utilization by an individual cell (MT^{-1})
X_a = concentration of active cells ($M_x L^{-3}$)
S_f = concentration of rate-limiting substrate (ML^{-3}) in contact with the cell
k = maximum specific rate of substrate utilization ($MM_x^{-1}T^{-1}$)
K_s = concentration at which the specific rate is one-half of k (ML^{-3})
V = volume containing cells (L^3)

The key parameters characterizing the kinetics of utilization of a substrate are k and K_s. A large value of k and a small value of K_s are associated with a rapid rate of biodegradation.

The kinetic parameters are not well known for many of the organic chemicals that commonly pollute ground water. The K_s parameter seems to vary widely (e.g., from as low as about 1 μg/l to hundreds of milligrams per liter).

Within the biofilm, substrates must be transported by molecular diffusion if they are to penetrate beyond the outer surface of the biofilm or microcolony. Diffusion is described by Fick's second law,

$$r_{\text{diff}} = -D_f \frac{d^2 S_f}{dz^2} V, \tag{4.7}$$

where

r_{diff} = rate of substrate accumulation due to diffusion (MT^{-1})
D_f = molecular diffusion coefficient of the substrate in the film or colony (L^2T^{-1})
z = distance dimension normal to the surface of the film or colony (L)

The simultaneous utilization and diffusion of substrate within the film or colony are usually represented as a steady-state mass balance on S_f:

$$\frac{d^2 S_f}{dz^2} = \frac{kX_a}{D_f} \frac{S_f}{K_s + S_f}. \tag{4.8}$$

Because the microorganisms are attached to a surface, substrates must be transported to the surface. This external mass transport is represented in the conventional manner by

$$J - K_m(S - S_s), \tag{4.9}$$

where

S = substrate concentration in the pore liquid (ML^{-3})
S_s = substrate concentration at the interface between the liquid and the biofilm or colony surface (ML^{-3})

Equations (4.8) and (4.9) are the ones that must be solved simultaneously to give J (see [4.5] and Rittmann and McCarty, 1981). Provided the amount of attached biomass is known, equation (4.5) and a solution for J can be employed for any type of rate-limiting substrate.

The microorganisms must be grown and sustained. At a minimum, they must consume an electron donor and an electron acceptor; nutrients (e.g., nitrogen and phosphorus) are also needed if cells are accumulating. One of these materials is "growth rate limiting" and must be modeled if the amount of active biomass is to be described. For attached biomass, the linkage of limiting-substrate utilization can be made by solving a mass balance equation on cell mass for a limiting electron donor, often called the primary substrate (Rittmann and McCarty, 1980; Saez and Rittmann, 1988). The mass balance (O) is given by

$$O = JY - b'X_f L_f, \tag{4.10}$$

where

Y = true yield of cell mass per unit of primary substrate consumed ($M_x M^{-1}$)
L_f = biofilm or microcolony thickness (L)
X_f = density of active cells in the biofilm or microcolony ($M_x L^{-3}$)
b' = overall biomass loss rate (T^{-1})

In summary, the rate term for biodegradation is given by (4.9). However, predicting J for a given compound requires knowledge of the amount of active biomass ($X_f L_f$) and the rate parameters (k, K_s, k_m, D_f) for that compound. Because the amount of active biomass depends on the utilization of the growth-rate-limiting substrate, that compound must be modeled. Often that limiting material is the electron donor or acceptor, but it need not be the compound of primary interest.

If the contaminant of interest is not the growth-rate-limiting substrate, its utilization does not affect the cell accumulation. The kinetics for non-growth-rate-limiting substrates fall into one of two classes, a nonlimiting necessary substrate and a secondary substrate.

A nonlimiting necessary substrate is an electron donor, electron acceptor, or nutrient that is required for cell growth or maintenance but is present at a concentration sufficiently high that it does not limit the overall rate of cell metabolism. The flux for such a material is proportional to J for the rate-limiting material times a stoichiometric ratio.

A secondary substrate is an organic compound whose utilization contributes negligible energy, electrons, or carbon for cell growth or maintenance. A secondary substrate contributes negligibly toward the accumulation of cells because of its low concentration, transient presence, or inability to support any growth. The last situation is known as co-metabolism. The rate of secondary-substrate utilization is determined by its intrinsic kinetic parameters, its concentration, and the amount of active biomass, which is controlled by the long-term availability of primary substrate.

Incorporation of Biological Processes into Solute Transport Models

Adding a biological reaction term, in the form of equation (4.9), to the solute mass balance presents no conceptual challenge to modeling subsurface transport. Because algorithms are available for computing J as a function of S and kinetic parameters, the biological reaction term can be treated as a pseudoconstant that is computed as needed by pseudoanalytical solutions (e.g., Rittmann and McCarty, 1981; Saez and Rittmann, 1988). This approach, using a pseudoconstant J, has been applied successfully many times in a research setting but is not yet common in field practice.

The application of the pseudoanalytical solutions can encounter four complications and practical problems. The first occurs when the system being modeled becomes large or spatially complicated. Then, the nonlinearity of the biological reaction terms (e.g., r_{bio} is not a first-order function of S) makes the computations very expensive if an accurate solution is to be attained. New techniques are needed for making tractable the solution of mass balance equations containing highly nonlinear reaction terms.

A second practical difficulty with modeling biological systems is that several components should be modeled. At a minimum, the active biomass needs to be estimated, but that task may require modeling the fate of one or more necessary substrates. If the compound of interest is not one of the necessary substrates, it needs to be modeled separately. The tools to model all of the components are available, but they must be combined properly. Clear distinctions must be made among primary substrates, nonlimiting necessary substrates, and secondary substrates. Again, all the tools have been properly combined for research investigations but are not being used routinely in practice.

A third complication is that increased accumulations of biomass in an aquifer can lead to loss of permeability, or clogging. The most obvious mechanism is growth of bacteria into the pores, thereby reducing the pore area available for flow. In addition, microbial action can reduce permeability through formation of gas pockets, precipitation of solids, or increase in the viscosity of the liquids from excretion of polymers. Incorporation of clogging into solute transport modeling is difficult for two reasons. First, the mechanisms of clogging are not yet well enough understood to allow formulation of quantitative expressions. Second, clogging of the pores alters the flow paths; thus water flow and solute transport models must become interactively coupled. Much research must still be done on all aspects of the clogging phenomenon.

A fourth complication involves substrates that are poorly soluble. Examples are organic solvents that form separate liquid phases, sorb strongly to solids, or volatilize to a gas phase. Incorporation of biodegradation into a model that already contains one or more transfers between phases is a very challenging problem. The main difficulties are two: (1) substrate mass balances are required in two or more phases, which intensifies computational demands, and (2) concentration gradients probably occur on a scale (e.g., micrometers to centimeters) much smaller than the model grid. The effect is to require a microscale in the direction normal to the phase interface. This microscale may force addition of another space dimension to the model, greatly increasing computational demands.

The four complications and practical problems can be accentuated when in situ bioremediation strategies are to be modeled. The addition and extraction of water through wells add to the local nonhomogeneities of flow velocity and solute concentration. The input of oxygen and nutrients is a hydraulics (delivery) problem that is often the limiting factor in the bioremediation strategy. The input of stimulating substrates or nutrients also induces significant and localized microbial growth, which can effect clogging. Hence, the modeling difficulties are made more intense by the localized and nonhomogeneous microbial activity created by bioreclamation practices. This adds to the problems already associated with heterogeneity in the permeability distribution.

TRANSPORT IN THE UNSATURATED ZONE

Reactive mass transport is of particular interest within the unsaturated zone. In cases where we are dealing with contaminants that are released at the soil surface (e.g., application of fertilizers and pesticides, land treatment of hazardous wastes, accidental spills of wastes, leaky storage tanks, lagoons or ponds used for storing waste liquids), the unsaturated zone may be thought of as a buffer zone that offers protection to the underlying aquifer. The unsaturated zone thickness may vary from a few to several hundred meters; water and vapors (and the contaminants dissolved in these fluids) must travel through the unsaturated zone and arrive in sufficient quantities at the water table to be an environmental or a health concern. These issues focus attention on the ability to quantify and model these processes in the unsaturated zone.

A large number of the processes, discussed earlier, will alter the nature and quantities of the contaminants arriving at the water table as a function of the travel time within the unsaturated zone. The unsaturated zone, particularly the top 1 or 2 m, is characterized by high microbial activity, which promotes biodegradation. This zone is also high in organic matter and clay content, which promotes sorption, biological degradation, and transformation. Of particular significance are the differences in the rates and magnitudes of these processes in the unsaturated zone as compared to the saturated zone.

The conventional view of mass transport in the unsaturated zone is simply that of advection moderated by both retardation and attenuation. Overall then, the presence of the unsaturated zone should in theory generally lead to decreased loadings of contaminants to the ground water. Vapor-phase transport, which occurs only in the unsaturated zone, can contribute to gaseous losses of the volatile contaminants. The unsaturated zone provides a buffer, either preventing or minimizing ground water contamination from chemicals applied at the ground surface. However, detection of a large number of volatile and nonvolatile contaminants (e.g., pesticides used widely in agriculture and organic contaminants derived from spills or leaks of gasoline and industrial solvents) in both shallow and deep ground water has raised questions as to the validity of what has been called the "filter fantasy," i.e., that the unsaturated zone acts as a protective buffer.

An alternative scenario is equally probable. The unsaturated zone might actually serve as a "source zone" for contamination of ground water. Pesticides and fertilizers sorbed on mineral and organic constituents of the solid matrix, as well as the residual amounts of gasoline or other organic solvents entrapped within the soil pores, may in fact be released slowly over a long time period, leading to long-term loadings of contaminants into the saturated zone. Thus short-term measures to remediate ground water, for example by pump-and-treat methods, may fail because of the long-term "bleeding" of contaminants from the unsaturated zone overlying the water table.

It is evident that the role of the unsaturated zone, either as a buffer or as a source, must be carefully evaluated in assessing ground water contamination and in selecting remedial measures. Coupling of simulation models developed for the unsaturated zone to those for the saturated zone is an essential element of ground water modeling and is necessary for devising appropriate management/regulatory policies.

The modeling of contaminant behavior in the unsaturated zone is designed generally to answer the following three questions, listed in order of priority and increasing complexity. First, when might a contaminant arrive at a specified depth? This requires a prediction of the *travel time* (t_r) for the contaminant to arrive at the specified depth (Z_c) of interest: for example, the bottom of the root zone, the bottom of the treatment zone at a hazardous waste land treatment (HWLT) facility, or the water table. Second, how much of the surface-applied (or spilled) contaminant might arrive at Z_c? This requires an estimation of the *mass loadings* (M_t) of the contaminant beyond the depth Z_c as influenced by retardation resulting from sorption and attenuation as a consequence of various biotic/abiotic transformations during contaminant transport through the unsaturated zone. Finally, it might also be necessary to predict the concentration distribution ($C[z,t]$) of the contaminant within the unsaturated zone such that the time changes in contaminant concentrations as well as fluxes ($J_c[t]$) at Z_c may be evaluated in addition to M_t. The spatial and temporal scales at which these questions need to be addressed and the ability to provide the necessary data characterizing the unsaturated zone and the contaminant determine the complexity of the model used and the reliability of the predictions provided by the model(s).

In answering the questions posed above, it is important to understand the coupling between the physical processes of *flow* and *storage*, the chemical processes of *retention* and *reaction*, and the biological processes of *degradation* (complete breakdown to nontoxic products) and *transformation* (partial decomposition that may or may not lead to the production of toxic by-products). It is also necessary to examine the differences in the rates and magnitudes of these processes as they occur in the unsaturated zone in contrast to what happens in the saturated zone.

As water infiltrates and redistributes within the unsaturated soil, various solutes dissolved in it are carried along. The advective and hydrodynamic transport of solutes is discussed elsewhere. Here it is sufficient to recognize that when the soil water flow is transient, solute transport is also transient. The advective velocity (v) at which a nonadsorbed solute is transported in an unsaturated soil is given by q/Θ (recall that both q and Θ vary with space and time during transient water flow). Thus the transport of a nonadsorbed, conservative solute (e.g., chloride) can be described, given only the knowledge of how water flows in a soil profile. For an adsorbed or nonconservative solute, however, retardation of transport, because of sorption and attenuation owing to transformations and reactions, must also be taken into account.

Because water flow in the unsaturated zone is episodic, so is contaminant transport. This feature is illustrated schematically in Figure 4.5 for vertical, downward leaching of nonsorbed and sorbed contaminants in a sandy soil as a result of rainfall over a 1-yr period. The progressive downward leaching, in a *stepwise* manner, of the contaminant pulse is clearly evident. Note that periods during which there is no contaminant transport (indicated by horizontal lines in Figure 4.5), even though rainfall occurs at the ground surface, are the periods when the soil-water depletion above the contaminant pulse, because of evapotranspiration losses, was not overcome by a given sequence of water input events. Further downward leaching of the contaminant (indicated by short vertical lines in Figure 4.5) can occur only when this soil-water deficit is overcome.

Attenuation is defined for the present discussion as the decrease in the total amount of the contaminant present within the unsaturated zone. Attenuation therefore includes all losses via various transformations but does not include the decrease in the contaminant concentration (i.e., *dilution*) resulting from hydrodynamic dispersion. Near the ground surface, where microbial activity is likely to be highest, losses due to microbial degradation will be the largest. Microbial activity declines rapidly with increasing soil depth, and losses are primarily due to chemical transformations (e.g., hydrolysis). The residence time within this biologically active zone is of paramount importance in determining the extent of attenuation and, hence, mass loadings to ground water. For the hypothetical case presented in Figure 4.5, the attenuation of three contaminants with time as they travel through the unsaturated zone is depicted in Figure 4.6. Note the changes in the slope of the decay curve, each change being coincidental with the movement of the contaminant pulse from a soil horizon of high microbial activity (faster rate of attenuation) to a deeper horizon with lower microbial activity and, hence, slower attenuation.

With this consideration, the significance of the episodic nature of contaminant transport within the unsaturated zone now becomes more apparent. The pattern of water input at the ground surface

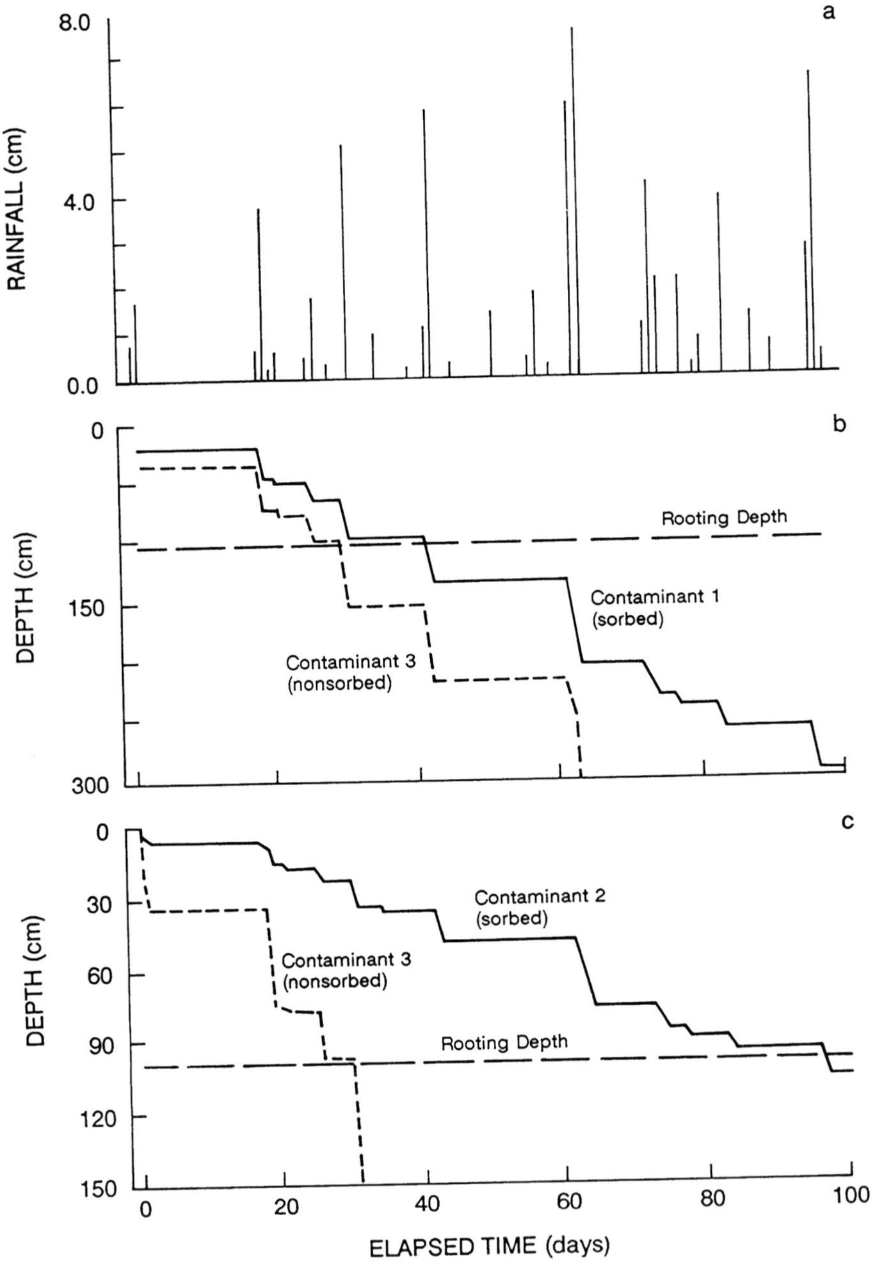

FIGURE 4.5 Sequential leaching of two sorbed contaminants (1,2) and one nonsorbed contaminant (3) through the root zone as a result of rainfall shown in (a). Note that contaminant 2 is sorbed to a greater extent than contaminant 1 and, as a consequence, is leached to a lesser extent.

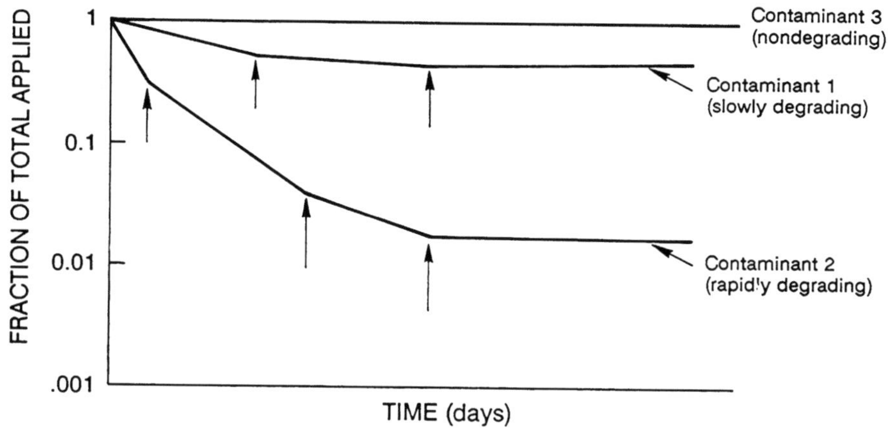

FIGURE 4.6 Attenuation of three contaminants during their transit through the unsaturated zone. The shifts in attenuation rates, indicated by arrows, coincide with leaching from one soil horizon to the next. Note the logarithmic scale used to show the amount of contaminant remaining in the unsaturated zone.

and the soil's physical properties control the temporal and depth variations in soil water flux which, in turn, dictates the residence time of the contaminant in each depth increment. Both retardation and the rate of attenuation vary with soil depth; therefore the extent of attenuation occurring within a given zone is dependent on the residence time in that zone. The actual mass loadings of a contaminant beyond the root zone (or to the water table) are also episodic. An example, based on simulations using an unsaturated zone model called PRZM, is shown in Figure 4.7 for the loadings of the nematocide aldicarb to shallow ground water beneath citrus groves. Note the variations in timing and amount of daily pesticide inputs into the shallow water table. These episodes (or loading events) are controlled by the dynamics of unsaturated water flow in the citrus root zone. Such model outputs are then used as inputs to a model that simulates pesticide behavior in the saturated zone (see Figure 4.8).

Unsaturated Flow and Transport in Structured Soils

The classical Richards equation for transient water flow and the advective-dispersive solute transport model may be adequate in homogeneous soils, but may not be appropriate for describing flow and transport in structured soils (van Genuchten, 1987). Structured

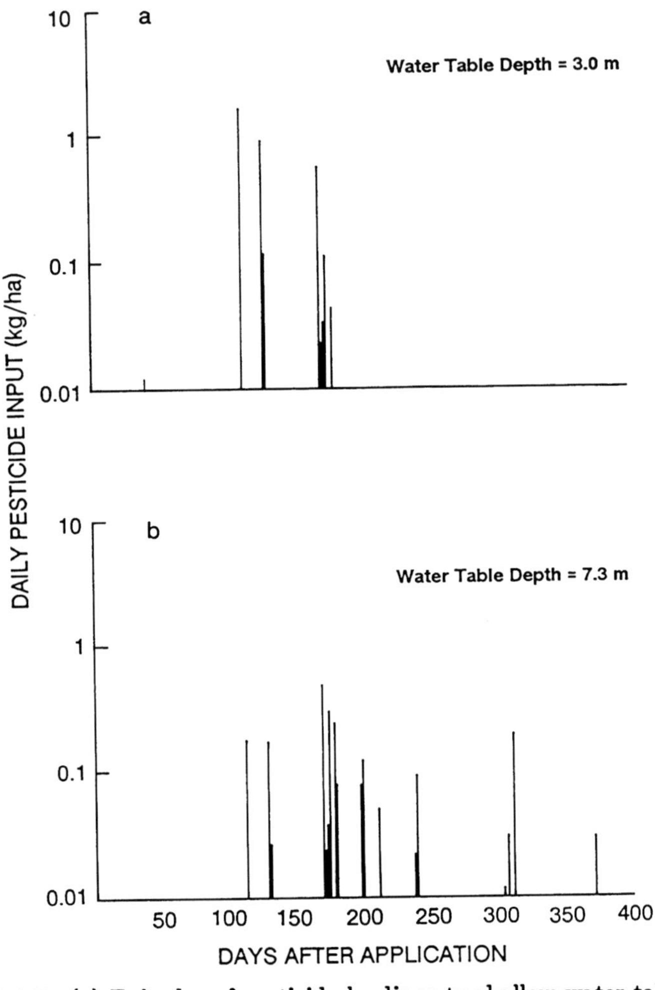

FIGURE 4.7 (a) Episodes of pesticide loadings to shallow water table located beneath a citrus grove. (b) Daily pesticide loadings predicted by PRZM and used in saturated zones simulations. SOURCE: Jones et al., 1987.

soils are characterized by large, more or less continuous voids often referred to as macropores (Luxmoore, 1981). Voids in porous media in which water is essentially not subjected to capillarity (capillary potential greater than −0.1 kPa) and that therefore may be wider than about 3 mm have been defined as macropores (Germann and Beven, 1981). A few examples of such macropores are interaggregate pores; interpedal voids; earthworm or gopher holes; decayed-root channels; and drying cracks and fissures in clay soils. It must be noted that designation of macropores based on size only is still disputed (see Germann, 1989).

During an infiltration event, water and solutes can preferentially flow into and through these macropores and bypass a major portion of the soil matrix. In this regard, the conceptual problem of predicting macropore flow is somewhat similar to that of describing solute transport in saturated, fissured and fractured media. In both cases, rapid flow in the macropores (or fissures) is accompanied by much slower infiltration or diffusion-controlled mass transfer into the soil matrix. The major distinction is, of course, that unlike the case for fissures and fractures in aquifers, flow and transport in macropores occur only when specific conditions are satisfied. Identification of these conditions and modeling of macropore flow and transport have been the main focus of recent investigations by soil scientists and subsurface hydrologists. The occurrence of macropore flow is determined by, among other factors, the antecedent soil-water conditions, hydraulic properties of the soil matrix, the rate of water input at the soil surface, and the spatial distribution (i.e., density) and interconnectedness of the macropore sequences. The impact of such preferential flow on solute transport is further determined by the rate of diffusive mass transfer into the soil matrix and the sorptive properties of the macropore and matrix regions.

While the impacts of preferential water flow on subsurface hydrology have been more thoroughly investigated, only recently have efforts been initiated to investigate the influence of macropore flow on solute transport in structured soils. Beven and Germann (1982) and White (1985) have reviewed the available experimental evidence for preferential flow and bypassing. One major impact of macropore flow is that of accelerated movement of surface-applied solutes (e.g., fertilizers, pesticides, and salts) through the vadose zone. Macropore flow is probably responsible for the frequent reports that field-measured dispersion coefficients are much larger (by an order of magnitude or more) than those measured in packed laboratory columns.

Germann (1989) summarized the attempts to model transient flow and transport in structured soils and grouped them into three basic approaches: (1) macroscopic averaging of flow and transport based on the "mobile-immobile" zone concept; (2) flow and transport based on various routing procedures along presumed stream lines; and (3) transfer function models based on a continuous velocity distribution. One of the major limitations in predicting macropore flow

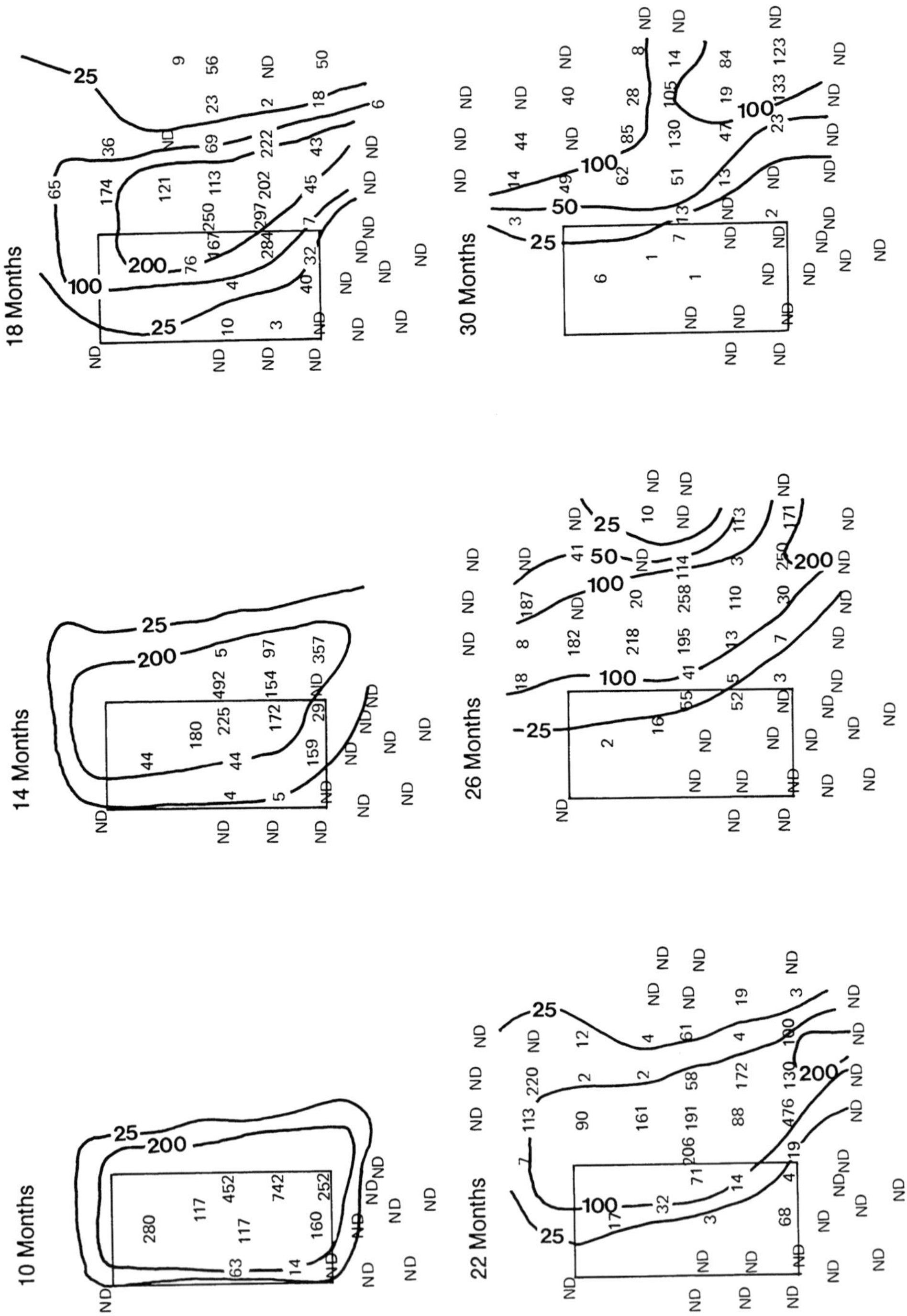
10 Months
14 Months
18 Months
22 Months
26 Months
30 Months

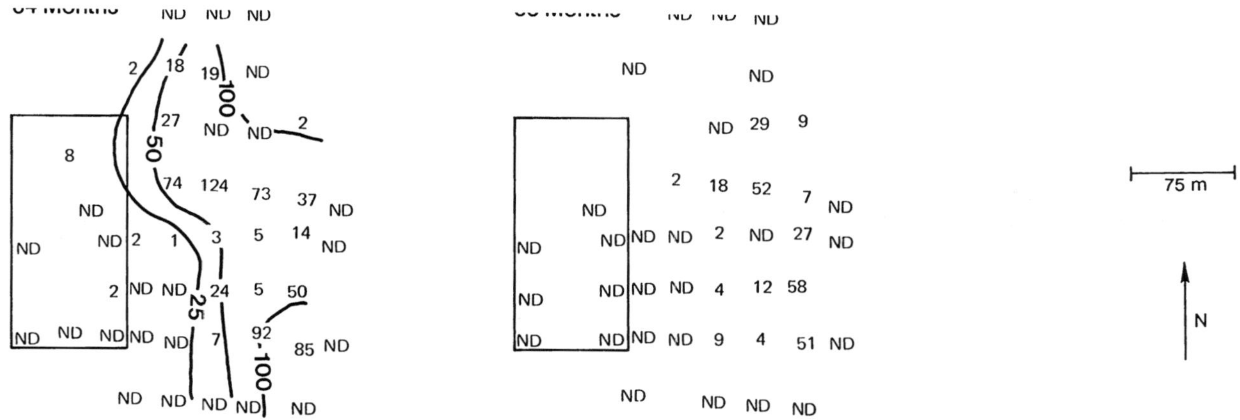

FIGURE 4.8 Measured and simulated movement of aldicarb residues. (Numbers represent maximum concentration measured in a well cluster, and contours show the simulation results if a half-life of 8 months is assumed; simulated results are not available after 34 months.) SOURCE: Jones et al., 1987.

and bypassing is perhaps not our inability to develop comprehensive mathematical models, but a limitation in characterizing the geometric and hydraulic features (e.g., size, length, spatial distributions, permeability, and interconnectedness) of the macropores and in providing the values for the required model parameters. In this sense, the problems of a ground water hydrologist dealing with flow and transport in fractured media are like those of a soil scientist. The added complexities faced by soil scientists are those of transient flow domain and uncertainty as to when macropore flow is dominant and how it should be modeled.

MULTIPHASE TRANSPORT

Two classes of multiphase flow and contaminant transport arise most often in ground water studies: seawater intrusion and organic fluid migration. The standard conceptual model of the vadose zone also includes two fluid phases—air and water—but this case has been treated previously. Codes that simulate the two classes of problems exist but have not been so heavily involved in water quality regulation or litigation as standard ground water flow and miscible solute transport codes. This is for three reasons: (1) the applications are primarily concerned with the resource (e.g., water or oil) quantity, (2) the technology is new and relatively untested, and (3) insufficient data exist to employ multiphase principles. In the following paragraphs, an outline of typical problems governing equations, necessary data, and typical codes is presented.

Seawater Intrusion

Freshwater supplies located adjacent to bodies of saltwater can be affected by water use strategies adopted by the public and industry. A typical setting for seawater intrusion is shown in Figure 4.9. Other settings can be envisioned, e.g., a confined aquifer instead of an unconfined aquifer, and saline contamination that completely underlies an overlying freshwater body. Clearly, the degree to which the freshwater and saltwater interface is dispersed could be important to a study. The extent to which the interface is smeared longitudinally may also be important in deciding upon the conceptual model. The proper choice of a conceptual model and method of analysis will be determined by the operative physics of the system and the behavior of interest in the study (Voss, 1984). If the area considered is large,

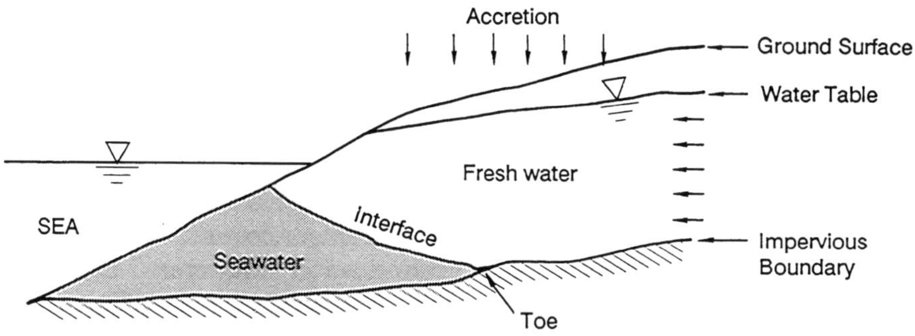

FIGURE 4.9 Seawater intrusion in an unconfined aquifer. SOURCE: After Sa da Costa and Wilson, 1979.

then the problem scale is such that an interface approximation is valid and saltwater and fresh water may be treated as immiscible.

The analysis of aquifers containing both fresh water and saltwater may be based on a variety of conceptual models (Voss, 1984). The range of numerical models includes dispersed interface models of either cross-sectional or fully three-dimensional fluid-density-dependent flow and solute transport simulation. Sharp interface models are also available for cross-sectional or areal applications. Of the sharp interface models, some account for the movement of both fresh water and saltwater, while others account only for freshwater movement. The latter models are based on an assumption of instantaneous hydrostatic equilibrium in the saltwater environment.

In the majority of cases involving seawater intrusion, water quality is viewed as good or bad; either it is fresh water or it is not a resource. Therefore many studies seek to determine acceptable levels of pumping or appropriate remediation or protection strategies. These resource management questions are resolved through fluid flow simulation and do not require solute transport simulations.

Organic Fluid Contamination

The migration and fate of organic compounds in the subsurface are of significant interest because of the potential health effects of these compounds at relatively low concentrations. A significant body of work exists within the petroleum industry regarding the movement of organic compounds, e.g., oil and gas resources. However, this capability has been developed for estimating resource recovery or production and not contaminant migration. To compound problems, the petroleum industry's computational capability is largely proprietary and is oriented toward deep geologic systems, which typically have higher temperatures and higher pressure environments than those encountered in shallow contamination problems.

Within the past decade, a considerable effort has been made to establish a capability to simulate immiscible and miscible organic compound contamination of ground water resources. Migration patterns associated with immiscible and miscible organic fluids are schematically described by Schwille (1984) and Abriola (1984). Figure 4.10 depicts one possible organic liquid contamination event. If not remediated, the migration of an immiscible organic liquid phase is of interest because it could represent an acute or chronic source of pollution. Movement of the organic liquid through the vadose zone is governed by the potential of the organic liquid, which in turn depends upon the fluid retention and relative permeability properties of the air/organic/water/solid system. As an organic liquid flows through a porous medium, some is adsorbed to the medium or trapped within the pore space. Specific retention defines that fraction of the pore space that will be occupied by organic liquid after drainage of the bulk organic liquid from the soil column. This organic contamination held within the soil column by capillary forces (at its residual saturation) represents a chronic source of pollution because it can be leached by percolating soil moisture and carried to the water table.

If the organic liquid is lighter than water, it may migrate as a distinct immiscible contaminant (the acute source) within the capillary fringe overlying the water table. The soluble fraction of the organic liquid will also contaminate the water table aquifer and migrate as a miscible phase within ground water. This is the situation shown in Figure 4.10. If the organic liquid is heavier than water, it will migrate vertically through the vadose zone and water table to directly contaminate the ground water aquifer. It may also penetrate water-confining strata that are permeable to the organic liquid and, consequently, contaminate underlying confined aquifers. The organic contaminant may form a pool on the bedrock of the aquifer and move in a direction defined by the bedrock relief rather than by the hydraulic gradient. Contamination of ground water occurs by dissolution of the soluble fraction into ground water contacting either the main body of the contaminant or the organic liquid held by specific retention within the porous medium.

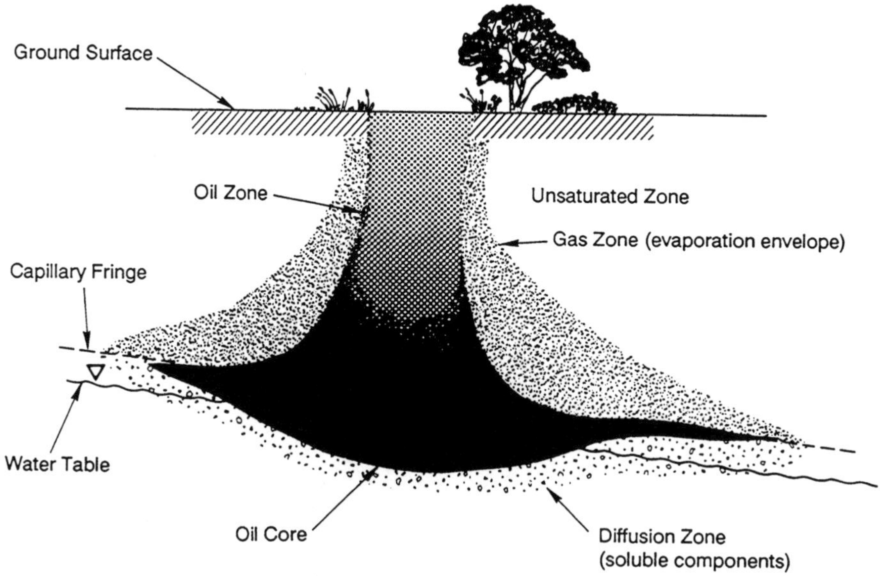

Figure 4.10 Organic liquid contamination of unsaturated and saturated porous media. SOURCE: After Abriola, 1984.

Governing Equations for Multiphase Flow

The region of greatest interest in seawater intrusion problems is the front between fresh water and seawater. The problem of salinity as a miscible contaminant in ground water is addressed with standard solute transport models. In reality, seawater is miscible with fresh water, and the front between the two bodies of water is really a transition zone. The density and salinity of water across the zone gradually vary from those of fresh water to those of seawater (Bear, 1979). A sharp or abrupt interface is assumed if the width of the transition zone is relatively small. Fresh water is buoyant and will float above seawater. The balance struck among fresh water (i.e., ground water) moving toward the sea, seawater contaminating the approaching fresh water by miscible displacement, and fresh water overlying seawater results in a nearly stationary saline wedge. Figure 4.9 illustrates the stationary saline wedge conceptual model. This wedge will change if influenced by pumping or changes in recharge.

While one can pose and solve the seawater intrusion problem as a single fluid having variable density (e.g., Segol et al., 1975; Voss, 1984), the most common approach has been to simulate fresh water and seawater as distinct liquids separated by an abrupt interface. Along the interface, the pressures of both liquids must be identical. Sharp interface methods are applied to both vertical cross sections (e.g., Volker and Rushton, 1982) and areal models (e.g., Sa da Costa and Wilson, 1979). The equations used to formulate the problem are the same as those used for the standard ground water flow problem. The only differences are that two equations are used (i.e., freshwater and seawater versions) and that their joint solution is conditioned to the pressure along the interface. Assuming that the response of the seawater domain is instantaneous and hence that hydrostatic equilibrium exists in the seawater domain, one can model the intrusion problem with a standard transient ground water flow model (Voss, 1984).

Pinder and Abriola (1986) provide a broad overview of the problem of modeling multiphase organic compounds in the subsurface. Abriola (1984) grouped models of multiphase flow and transport into two categories, those that address the migration of a miscible contaminant in ground water and those that address two or more distinct liquid phases. The former category of models addresses the far-field problem of chronic miscible contamination. Standard ground water flow and solute transport codes can be applied to these organic compound contamination problems. However, standard codes may require modifications to address biodegradation or sorption characteristics of a specific organic compound.

As in the case of seawater intrusion, the region of greatest interest is the region exhibiting multiphase behavior. The problem of organic contamination is more complex for two reasons: (1) in general, a stationary interface will not exist, and (2) one is often interested in contamination of unsaturated soil deposits as a precursor to contamination of a ground water aquifer. Interest in the migration and fate of organic compounds has required that transient analysis methods be developed. Such methods enable one to simulate the movement of bulk contamination through the vadose zone and into a ground water aquifer. One is also able to estimate the mass of contamination held in the media by specific retention. Because the front is not stationary, one must model liquid/solid interactions that govern the movement of each fluid in the presence of others. The equations describing multiphase flow and transport are similar to those previously described for simulating water movement and solute transport in unsaturated soils. One fluid flow (e.g., fluid mass conservation) equation is required for each fluid phase simulated (e.g., gas, organic liquid, water). Rather than simulate distinct fluid regions separated by abrupt interfaces, one simulates a continuum shared by each of the fluids of interest. The equation set is coupled by the fluid retention and relative permeability relationships of the multiphase system.

Miscible displacement of trace quantities of an organic fluid can occur within the water and gas phases. This is a common occurrence; however, it greatly complicates the mass balance equation for the organic fluid. The statement of mass conservation must now account for organic mass entrained in the water and gas phases as well as the organic mass held in the immiscible fluid phase. Transport processes are introduced into the conservation equations, and the exchange of organic mass between fluid phases must be accounted for through partition coefficients.

Abriola (1988) and Allen (1985) review models available for the simulation of multiphase problems. A variety of solutions have been published for multiphase contamination problems. This is due to the complexity of the overall problem and the variety of approaches that can be taken to provide an approximate solution. A useful hierarchy of modeling approaches is as follows: sharp interface approximations, immiscible phase flow models incorporating capillarity, and compositional models incorporating interphase transfer.

Examples of models based on sharp interface approximations are those of Hochmuth and Sunada (1985), Schiegg (1986), and van Dam (1967). Immiscible phase models incorporating capillarity allow a more realistic simulation of the specific retention phenomena but do not address hysteresis in the fluid-soil interaction. Examples of these models are presented by Faust (1985), Kuppusamy et al. (1987), and Osborne and Sykes (1986). Compositional models incorporating interphase transfer are extremely complex and require the most data, many of which are not routinely available for contaminants of interest. Examples of these models are presented by Abriola and Pinder (1985a,b), Baehr and Corapcioglu (1987), and Corapcioglu and Baehr (1987).

Parameters and Initial and Boundary Conditions for Multiphase Flow

The physical complexity exhibited by multiphase flow models consumes all available computer resources. This strain on computer resources has precluded acknowledgment, in models, of the complexities of heterogeneous media that are spatially distributed in the real environment. At the present time, computational resources restrict fully three-dimensional problems to homogeneous, porous media. Realistically, currently available computational resources are best suited to address conceptual models.

Model parameters necessary for the simulation of seawater intrusion are basically identical to those required for the simulation of ground water flow; however, two-fluid models require duplicate parameters for fresh water and saltwater. A great many more model parameters are necessary for a complete analysis of immiscible organic contaminant migration in the subsurface. While the seawater intrusion problem is restricted to saturated porous media, organic fluid migration often occurs in the unsaturated zone. Consequently, fluid retention and relative permeability properties are required for the air/organic/water/solid system. Other standard data requirements for multiphase fluid flow simulation include porosity, compressibility of liquids and porous media (or storage coefficient), fluid densities and viscosities, and the intrinsic permeability tensor.

As in the case of the fluid flow simulations, model parameters for transport simulations are more detailed for the organic fluid migration problem than for the seawater intrusion problem. Model parameter requirements for solute migration within variable-density seawater intrusion are very similar to the requirements of any single-phase saturated zone model; however, duplicate data sets are required for freshwater and seawater domains. Parameters necessary for detailed analysis of organic liquid transport phenomena include macroscopic diffusion and dispersion coefficients for each fluid phase (e.g., gas, water, or organic liquid), partition coefficients for water-gas and water-organic phases, sorption model parameters for alternative sorption models, and degradation model parameters for the organic fluid.

Certainly, the more complex and complete models of multiphase contaminant problems require more data. If one considers only the immiscible flow problem in an attempt to estimate the migration of the bulk organic plume, then one will not require any of the miscible displacement (transport) parameters. If one assumes that the gas phase is static, one greatly reduces the data requirement in terms of both flow and transport phenomena. Key data for any analysis of multiphase migration are the fluid retention and relative permeability characteristics for the fluids and media of interest. The media porosity and intrinsic permeability, as well as fluid densities and viscosities, are also essential.

All comments made regarding boundary and initial conditions for flow and transport of a single-phase contamination analysis also apply to a multiphase analysis. Aspects of transient analysis can be important in seawater intrusion problems because of seasonal pumping stresses. Transient analyses are also essential for organic fluid migration simulation because of interest in the migration and fate of these potentially harmful substances.

Spatial dimensionality of a multiphase analysis can influence results. For example, in the real, fully three-dimensional environment, a heavier-than-water organic fluid can move vertically through the soil profile and form a continuous distinct fluid phase from the water table to an underlying impermeable medium. Ground water will simply move around the immiscible organic fluid as though it were an impermeable object. Attempts to analyze such a situation in a vertical cross section with a two-dimensional multiphase model will fail because the organic fluid will act as a dam to laterally moving ground water. Thus only an intermittent source of immiscible organic fluid can be analyzed. Note that such an analysis will be flawed for most real-world applications because it will represent a laterally infinite intermittent source rather than a point source of pollution.

Problems Associated with Multiphase Flow

The problems associated with modeling multiphase flow include the following:

- magnitude of computational resources required to address all complexities;
- data requirements of the multiphase problem that are independent of consideration of spatial variability, paucity of data specific to soils and organic contaminants of interest, and no way to address the problem of mixtures of organics;
- absence of hysteresis submodels needed to address retention capacity of porous media and to enable one to simulate purging of the environment;
- virtual omission of any realistic surface geochemistry or microbiology submodels necessary to more completely describe the assimilative or attenuative capacity of the subsurface environment; and
- viscous fingering and its relationship to spatial variability occurring in the natural environment.

REFERENCES

Abriola, L. M. 1984. Multiphase Migration of Organic Compounds in a Porous Medium: A Mathematical Model, Lecture Notes in Engineering, Vol. 8. Springer-Verlag, Berlin.

Abriola, L. M. 1988. Multiphase Flow and Transport Models for Organic Chemicals: A Review and Assessment. EA-5976, Electric Power Research Institute, Palo Alto, Calif.

Abriola, L. M., and G. F. Pinder. 1985a. A multiphase approach to the modeling of porous media contamination by organic compounds, 1. Equation development. Water Resources Research 21(1), 11–18.

Abriola, L. M., and G. F. Pinder. 1985b. A multiphase approach to the modeling of porous media contamination by organic compounds, 2. Numerical simulation, Water Resources Research 21(1), 19–26.

Allen, III, M. B. 1985. Numerical modeling of multiphase flow in porous media. In Proceedings, NATO Advanced Study Institute on Fundamentals of Transport Phenomena in Porous Media, July 14–23, J. Bear and M. Y. Corapcioglu, eds. Martinus Nijhoff, Newark, Del.

Anderson, M. P. 1984. Movement of contaminants in groundwater: Groundwater transport—Advection and dispersion. Pp. 37–45 in Groundwater Contamination. National Academy Press, Washington, D.C.

Baehr, A. L., and M. Y. Corapcioglu. 1987. A compositional multiphase model for ground water contamination by petroleum products, 2. Numerical solution. Water Resources Research 23(1), 201–213.

Bear, J. 1979. Hydraulics of Groundwater. McGraw-Hill, New York, 567 pp.

Beven, K., and P. F. Germann. 1982. Macropores and water flows in soils. Water Resources Research 18, 1311–1325.

Cederberg, G. A., R. L. Street, and J. O. Leckie. 1985. A groundwater mass-transport and equilibrium chemistry model for multicomponent systems. Water Resources Research 21(8), 1095–1104.

Corapcioglu, M. Y., and A. L. Baehr. 1987. A compositional multiphase model for groundwater contamination by petroleum products, 1. Theoretical considerations. Water Resources Research 23(1), 191–200.

Davis, A. D. 1986. Deterministic modeling of a dispersion in heterogeneous permeable media. Ground Water 24(5), 609–615.

Delany, J. M., I. Puigdomenech, and T. J. Wolery. 1986. Precipitation kinetics option of the EQ6 Geochemical Reaction Path Code. Lawrence Livermore National Laboratory Report UCRL-53642, Livermore, Calif. 44 pp.

Domenico, P. A., and G. A. Robbins. 1984. A dispersion scale effect in model calibrations and field tracer experiments. Journal of Hydrology 70, 123–132.

Faust, C. R. 1985. Transport of immiscible fluids within and below the unsaturated zone: A numerical model. Water Resources Research 21(4), 587–596.

Felmy, A. R., S. M. Brown, Y. Onishi, S. B. Yabusaki, R. S. Argo, D. C. Girvin, and E. A. Jenne. 1984. Modeling the transport, speciation, and fate of heavy metals in aquatic systems. EPA Project Summary. EPA-600/53-84-033, U.S. Environmental Protection Agency, Athens, Ga., 4 pp.

Gelhar, L. W. 1986. Stochastic subsurface hydrology from theory to applications. Water Resources Research 22(9), 135s–145s.

Germann, P. F. 1989. Approaches to rapid and far-reaching hydrologic processes in the vadose zone. Journal of Contamination Hydrology 3, 115–127.

Germann, P. F., and K. Beven. 1981. Water flow in soil macropores, 1, An experimental approach. Journal of Soil Science 32, 1–13.

Goode, D. J., and L. F. Konikow. 1988. Can transient flow cause apparent transverse dispersion? (abst.). Eos, Transactions of the American Geophysical Union 69(44), 1184–1185.

Hochmuth, D. P., and D. K. Sunada. 1985. Ground-water model of two-phase immiscible flow in coarse material. Ground Water 23(5), 617–626.

Hostetler, C. J., R. L. Erikson, J. S. Fruchter, and C. T. Kincaid. 1988. Overview of FASTCHEM^{TM} Code Package: Application to Chemical Transport Problems, Report EQ-5870-CCM, Vol. 1. Electric Power Research Institute, Palo Alto, Calif.

Jones, R. L., A. G. Hornsby, P. S. C. Rao, and M. P. Anderson. 1987. Movement and degradation of aldicarb residues in the saturated zone under citrus groves on the Florida ridge. Journal of Contaminant Hydrology 1, 265–285.

Konikow, L. F. 1981. Role of numerical simulation in analysis of groundwater quality problems. Pp. 299–312 in The Science of the Total Environment, Vol. 21. Elsevier Science Publishers, Amsterdam.

Konikow, L. F. 1988. Present limitations and perspectives on modeling pollution problems in aquifers. Pp. 643–664 in Groundwater Flow and Quality Modelling, E. Custudio, A. Gurgui, and J. P. Lobo Ferreira, eds. D. Reidel, Dordrecht, The Netherlands.

Konikow, L. F., and J. M. Mercer. 1988. Groundwater flow and transport modeling. Journal of Hydrology 100(2), 379–409.

Kuppusamy, T., J. Sheng, J. C. Parker, and R. J. Lenhard. 1987. Finite-element analysis of multiphase immiscible flow through soils. Water Resources Research 23(4), 625–631.

Lindberg, R. D., and D. D. Runnells. 1984. Groundwater redox reactions: An analysis of equilibrium state applied to Eh measurements and geochemical modeling. Science 225, 925–927.

Luxmoore, R. J. 1981. Micro-, meso- and macro-porosity of soil. Soil Science Society of America Journal 45, 671.

Mercer, J. M., L. R. Silka, and C. R. Faust. 1983. Modeling ground-water flow at Love Canal, New York. Journal of Environmental Engineering ASCE 109(4), 924–942.

Miller, D., and L. Benson. 1983. Simulation of solute transport in a chemically reactive heterogeneous system: Model development and application. Water Resources Research 19, 381–391.

Naymik, T. G. 1987. Mathematical modeling of solute transport in the subsurface. Critical Reviews in Environmental Control 17(3), 229–251.

Osborne, M., and J. Sykes. 1986. Numerical modeling of immiscible organic transport at the Hyde Park landfill. Water Resources Research 22(1), 25–33.

Parkhurst, D. L., D. C. Thorstenson, and L. N. Plummer. 1980. PHREEQE—A computer program for geochemical calculations. U.S. Geological Survey Water Resources Investigation 80-96, 210 pp.

Peterson, S. R., C. J. Hostetler, W. J. Deutsch, and C. E. Cowan. 1987. MINTEQ User's Manual. Report NUREG/CR-4808, PNL-6106, Prepared by Battelle Pacific Northwest Laboratory for U.S. Nuclear Regulatory

Commission, Washington, D.C., 148 pp. (available from National Technical Information Service, U.S. Department of Commerce, Springfield VA 22161).

Pinder, G. F., and L. M. Abriola. 1986. On the simulation of nonaqueous phase organic compounds in the subsurface. Water Resources Research 22(9), 109s–119s.

Pinder, G. F., and A. Shapiro. 1979. A new collocation method for the solution of the convection-dominated transport equation. Water Resources Research 15(5), 1177–1182.

Plummer, L. N., B. F. Jones, and A. H. Truesdell. 1976. WATEQF—A FORTRAN IV version of WATEQ, a computer code for calculating chemical equilibria of natural waters. U.S. Geological Survey Water Resources Investigation 76-13, 61 pp.

Rittmann, B. E., and P. L. McCarty. 1980. Model of steady-state-biofilm kinetics. Biotechnology and Bioengineering 22, 2343–2357.

Rittmann, B. E., and P. L. McCarty. 1981. Substrate flux into biofilms of any thickness. Journal of Environmental Engineering 107, 831–849.

Rubin, J. 1983. Transport of reacting solutes in porous media: Relation between mathematical nature of problem formulation and chemical nature of reactions. Water Resources Research 19(5), 1231–1252.

Sa da Costa, A. A. G., and J. L. Wilson. 1979. A Numerical Model of Seawater Intrusion in Aquifers. Technical Report 247, Ralph M. Parsons Laboratory, Massachusetts Institute of Technology, Cambridge.

Saez, P. B., and B. E. Rittmann. 1988. An improved pseudo-analytical solution for steady-state-biofilm kinetics. Biotechnology and Bioengineering 32, 379–385.

Scheidegger, A. E. 1961. General theory of dispersion in porous media. Journal of Geophysical Research 66(10), 3273–3278.

Schiegg, H. O. 1986. 1.5 Ausbreitung von Mineralol als Fluessigkeit (Methode zur Abschaetzung). In Berteilung und Behandlung von Mineralolschadensfallen im Hinblick auf den Grundwasserschutz, Teil 1, Die wissenschaftlichen Grundlagen zum Verstandnis des Verhaltens von Mineralol im Untergrund. LTwS-Nr. 20. Umweltbundesamt, Berlin. [Spreading of Oil as a Liquid (Estimation Method). Section 1.5 in Evaluation and Treatment of Cases of Oil Damage with Regard to Groundwater Protection, Part 1, Scientific Fundamental Principles for Understanding the Behavior of Oil in the Ground. LTwS-Nr. 20. Federal Office of the Environment, Berlin.)

Schwille, F. 1984. Migration of organic fluids immiscible with water in the unsaturated zone. Pp. 27–48 in Pollutants in Porous Media, The Unsaturated Zone Between Soil Surface and Groundwater, B. Yaron, G. Dagan, and J. Goldshmid, eds. Ecological Studies Vol. 47, Springer-Verlag, Berlin.

Segol, G., G. F. Pinder, and W. G. Gray. 1975. A Galerkin-finite element technique for calculating the transient position of the saltwater front. Water Resources Research 11(2), 343–347.

Smith, L., and F. W. Schwartz. 1980. Mass transport, 1, A stochastic analysis of macroscopic dispersion. Water Resources Research 16(2), 303–313.

Sposito, G., and S. V. Mattigod. 1980. GEOCHEM: A Computer Program for the Calculation of Chemical Equilibria in Soil Solutions and Other Natural Water Systems. Department of Soils and Environment Report, University of California, Riverside, 92 pp.

van Dam, J. 1967. The migration of hydrocarbons in a water-bearing stratum. Pp. 55–96 in The Joint Problems of the Oil and Water Industries, P. Hepple, ed. The Institute of Petroleum, 61 New Cavendish Street, London.

van der Heijde, P. K. M., Y. Bachmat, J. D. Bredehoeft, B. Andrews, D. Holtz, and S. Sebastian. 1985. Groundwater management: The use of numerical models. Water Resources Monograph 5, 2nd ed. American Geophysical Union, Washington, D.C., 180 pp.

van Genuchten, M. Th. 1987. Progress in unsaturated flow and transport modeling. U.S. National Report, International Union of Geodesy and Geophysics, Reviews of Geophysics 25(2), 135–140.

Volker, R. E., and K. R. Rushton. 1982. An assessment of the importance of some parameters for sea-water intrusion in aquifers and a comparison of dispersive and sharp-interface modeling approaches. Journal of Hydrology 56(3/4), 239–250.

Voss, C. I. 1984. AQUIFEM-SALT: A Finite-Element Model for Aquifers Containing a Seawater Interface. Water-Resources Investigations Report 84-4263, U.S. Geological Survey, Reston, Va.

White, R. E. 1985. The influence of macropores on the transport of dissolved suspended matter through soil. Advances in Soil Science 3, 95–120.

Wolery, T. J., K. J. Jackson, W. L. Bourcier, C. J. Bruton, B. E. Viani, and J. M. Delany. 1988. The EQ3/6 software package for geochemical modeling: Current status. American Chemical Society, Division of Geochemistry, 196th ACS National Meeting, Los Angeles, Calif., Sept. 25–30 (abstract).

5
Experience With Contaminant Flow Models in the Regulatory System

INTRODUCTION

This chapter is divided into two parts: (1) a review of federal regulations and guidance concerning the use of contaminant transport models, and (2) five case studies illustrating the site-specific application of such models. These sections are based on the committee's review and interpretation of these regulations and guidance, existing reports on the use of such models, discussions with agency personnel, and the personal experience of the committee members.

This chapter focuses on the regulations, guidance, and practices of the U.S. Nuclear Regulatory Commission (USNRC) and the U.S. Environmental Protection Agency (EPA). These two regulatory agencies deal with contaminant transport from historic or proposed disposal facilities and recognize the need to evaluate present conditions and predict potential migrations. Both agencies have programs in place that require modeling. However, each agency suffers from unique problems that reflect its particular regulatory concerns.

The USNRC has had a number of years to prepare for an application for a high-level radioactive waste disposal. As a result, the agency has had the opportunity to develop detailed procedures on reviewing model applications. Unfortunately, because of changes in federal programs, the developed procedures are largely untested. In contrast, EPA has had to evaluate a large number of modeling studies as part of the Superfund program. Because of the rapid increase in sites being evaluated, EPA has not had an opportunity to develop a systematic plan for model review or application. In the following sections the two agencies' approaches to the use and review of models are summarized.

U.S. NUCLEAR REGULATORY COMMISSION REGULATIONS AND GUIDANCE

One of the USNRC's responsibilities is the licensing of facilities for the disposal of low-level and high-level radioactive wastes (see 10 CFR Part 61, "Licensing Requirements for Land Disposal of Radioactive Waste," and 10 CFR Part 60, "Disposal of High-Level Radioactive Wastes in Geologic Repositories; Licensing Procedures," respectively). To be licensed, a facility must meet certain requirements. For example, one requirement is that the site be capable of being modeled (10 CFR Part 61). Thus the USNRC has embedded into its regulations and guidance general principles concerning contaminant transport modeling. However, this guidance is largely untested because the USNRC has performed only limited licensing for waste disposal facilities.

Low-level radioactive waste (LLW) is generated by a number of institutions including industries, laboratories, hospitals, and facilities involved in the nuclear fuel cycle. Wastes are packaged and placed in shallow excavations or engineered structures that are then backfilled and capped to limit infiltration. The USNRC LLW disposal regulations specify performance objectives and specific technical requirements for site suitability that are designed to adequately protect public health (Siefken et al., 1982). One of the requirements is that "the disposal site shall be capable of being characterized, modeled, analyzed, and monitored" (U.S. Nuclear Regulatory Commission, 1987). The purpose of this requirement is to ensure that the hydrogeological conditions of the site are adequately understood through field studies.

The USNRC has also developed standard review plans (SRPs) (U.S. Nuclear Regulatory Commission, 1987) that direct the USNRC staff in evaluating the potential for migration for a disposal facility. Review plans have been issued to evaluate a number of potential migration pathways including radionuclide movement through the ground water and movement of radionuclides resulting from infiltration through the ground surface. The SRPs for ground water and infiltration contain information on the amount of modeling planned by the regulatory agency as well as the type of issues that will be reviewed by the agency.

The SRP indicates that the license application will be reviewed to determine whether the use of the input parameters has been justified and whether the data are sufficient to provide a reasonably accurate or conservative analysis regarding ground water pathways. The transport models will also be evaluated for their defensibility, suitability, and basic conservatism. The codes must be based on sound physical, chemical, and mathematical principles and must be correctly applied and sufficiently documented.

The applicant must supply the following:

- a complete description of the contaminant transport pathways between the engineered disposal unit and the site boundary and existing or known future ground water user locations;
- estimates and justification for the physical and chemical input parameters used in the transport models to calculate radionuclide concentrations;
- a description of the contaminant transport models used in the analysis, including modeling procedures and complete documentation of the codes as required in NUREG-0856 (U.S. Nuclear Regulatory Commission, 1987, p. 6.1.5.1-4);
- the justification, documentation, verification, and calibration of any equations or program codes used in the analyses; and
- the description of data and justification for the manipulation of any data used in the analyses (p. 6.1.5.2-3).

The SRP does not attempt to quantify the level of information required to adequately characterize the potential ground water transport at the sites nor does it outline the acceptance criteria for adequate site modeling. To evaluate the applicant's submittal, the USNRC will use "simple analytical modeling techniques with demonstrably conservative assumptions and coefficients" (p. 6.1.5.1-3). The SRP does not outline which codes will be used, and no other supporting documentation was provided that outlined the codes planned for use by the USNRC.

The SRP guidance states that "if the applicant's results are *more realistic than conservative*, then the applicant must clearly justify the application and results of the model" (p. 6.1.5.2-3). More sophisticated numerical modeling will be performed by the USNRC when the issues relating to the applicant's modeling studies cannot be resolved. The SRP does not discuss the apparent disparity between requesting field information to characterize the site and the use of conservative data in the modeling process.

The LLW program appears to have developed a systematic plan for incorporating modeling into the site evaluation process. The plan has attempted to consider, in a general way, the reliability of the data input, as well as the documentation and reliability of the computer codes. The program, as outlined in the USNRC guidance and the SRPs, appears to emphasize conservatism, although the regulations place equal emphasis on collecting adequate field information. Also, the program has not attempted to direct applicants toward particular computer codes because the codes the USNRC will use are not defined.

The USNRC also has published extensive documentation on the codes that are planned for potential use in evaluation of license applications. The publication of this documentation allows license applicants to consider using USNRC codes or to review their code choices against the USNRC-distributed tools.

The USNRC guidance is designed to evaluate whether the models accurately simulate the phenomena that are considered and to determine whether the numerical approximations accurately solve the mathematical equations. The test problems include analytical and semianalytical solutions, as well as problems based on laboratory or field studies.

By providing a standardized process of model evaluation, the USNRC is attempting to limit the amount of code comparison that will be required at the time of license application. The USNRC (1982) outlines the level of documentation deemed adequate, i.e.,

> [t]he documentation of mathematical models and numerical methods will provide the basis for USNRC's review of the theory and means of solution used in the code. It should contain derivations and justification for the model.

The documentation will help the USNRC in understanding modeling results that are submitted by the applicant during the licensing process and permits the USNRC to install and use the code on its own computer.

The USNRC guidance also outlines a computer software management system that will provide a software storage system to ensure future retrievability of computer codes and will provide a standardized testing process for applied codes. The storage system will include a catalog of modifications and the most updated version of the codes in use.

The USNRC is (1) assembling mathematical models for assessing Department of Energy (DOE) demonstrations; (2) developing computer software for use in assessing the long-term risk from disposal of radioactive wastes in deep geologic formations, in estimating dose commitments and potential adverse health effects from released radionuclides, and in performing sensitivity and uncertainty analyses; and (3) developing a quality assurance program to ensure adequate quality in computer codes developed and in data generated by these codes, as well as for maintenance of the programs. This program requires peer review and management approval to ensure a systematic record of calculations and analyses that are performed.

In summary, the USNRC has attempted to define a process that considers not only the problems in evaluating model results but also the issues surrounding code selection and application. The guidance documents have attempted to direct applicants to the appropriate level of code review without limiting the choice of code selection.

U.S. ENVIRONMENTAL PROTECTION AGENCY REGULATIONS AND GUIDANCE

The U.S. Environmental Protection Agency uses a wide variety of contaminant transport models and has a large number of specific sites where such models are used and will be used. The key EPA regulations and guidance affecting the use of contaminant transport models—e.g., those in the Superfund, hazardous waste management, and underground injection programs—are included in the following discussion.

Superfund

Law and Regulations

Superfund is the environmental law that authorizes EPA to

- identify sites where hazardous substances have been released into the environment;
- clean up such contamination and recover the costs from the responsible private parties; or

- in the alternative, (1) order a private party to perform the cleanup or (2) obtain a voluntary agreement from such private party (called potentially responsible party [PRP]) to perform the cleanup.[1]

Superfund is primarily directed at cleanup of inactive hazardous waste sites. Courts generally have resolved legal uncertainties and issues of statutory interpretation in favor of the government in order to hold private parties liable, because

> [g]iven the remedial nature of . . . [Superfund] its provisions should be afforded a broad and liberal construction so as to avoid frustration of prompt response efforts or so as to limit the liability of those responsible for clean-up costs beyond the limits expressly provided.[2]

If EPA performs the remedy with money from Superfund, the remedy is selected after a review of remedial alternatives. This process is subject to public comment. At other sites, EPA negotiates the remedy necessary for the site with the PRPs, such as in the S-Area landfill case (see Case Study 5, this chapter). These negotiated remedies are then incorporated into a consent decree (a legal document that resolves a lawsuit without a determination of liability, but requires the defendant to perform an action, e.g., installation of tile drains and a cap, and/or monetary payment).

Guidance

Modeling may be used in the Superfund program to (1) guide the placement of monitoring wells (Environmental Protection Agency, 1988b); (2) predict concentrations in ground water for an assessment of the present and future risks at the site (Environmental Protection Agency, 1986a, 1988b); (3) assess the feasibility and efficacy of remedial alternatives (Environmental Protection Agency, 1988b); (4) predict the concentration for an assessment of the residual risk after implementation of the preferred remedial action (Environmental Protection Agency, 1988b); or (5) apportion liability among responsible parties.

Contaminant transport modeling is important in the process of estimating exposure and therefore risk. Regardless of the toxicity of the chemical, no injury can occur unless there is exposure. The chemicals must migrate from the source of contamination to a point where they come into contact with humans and interact biologically with the human body. Modeling can be "used as a tool . . . to estimate plume movement . . ." (Environmental Protection Agency,

1988b). Models are most helpful when rough estimates are required.[3] A worst-case estimate (an estimate where all assumptions are chosen so as not to underestimate the possible exposure) may indicate that little risk exists if significant exposures are not predicted. However,

> [a]s more resources are devoted to an exposure assessment and more studies conducted, a refined assessment is generated. Often there will be several stages of refinement of an assessment, and the degree of refinement and accuracy finally required will be related to the certainty needed to enable risk management decisions [e.g., selecting a ground water cleanup level versus evaluating the most cost-effective method of achieving that level].[4]

The EPA *Superfund Public Health Evaluation Manual* guidance (Superfund guidance) specifies that realistic exposure assumptions based on the best data available should be used.[5] Superfund guidance requires EPA to consider systematically the extent of chemical fate and transport in each environmental medium in order to account for the behavior of all released chemicals (Environmental Protection Agency, 1986a, p. 39; see also 40 CFR §§300.68[e][1], 300.68[h][2][iv], 300.68[i][1]). A ground water concentration, based on such model estimation, is then compared to levels of public health concern, e.g., a drinking water standard or a risk-based cleanup level (Zamuda, 1986). EPA advises that "caution should be used when applying models at Superfund sites because there is uncertainty whenever subsurface movement is modeled, particularly when the results of the model are based on estimated parameters" (Environmental Protection Agency, 1988b, p. 3-22).

Superfund guidance provides a general framework for selecting and applying models (Environmental Protection Agency, 1988b, p. 3-33; 1988c). Superfund modeling guidance recognizes the potential problem posed by the large range of models available and attempts to support users by providing guidelines for model choice (Environmental Protection Agency, 1988c). These criteria allow users to more easily justify code choice during discussions with regulators and may provide some common ground for discussing the use of alternative codes.

Three types of criteria are recommended for use in model selection: objective, technical, and implementation. The objective criteria used relate to the level of modeling detail needed to meet the objectives of the study, i.e., (1) performing a screening study or (2) performing a detailed study (Environmental Protection Agency,

1988c). Because the purpose of a screening study would be to obtain a general understanding of site conditions or to make general comparisons between sites, a simple model may be suitable at that stage.

The technical criterion used for model selection relates to the model's ability to simulate site-specific transport and fate phenomena of interest at the site. There are three areas where technical criteria should be developed: transport and transformation processes, domain configuration, and fluid media properties.

The third type of criterion used for model selection relates to the ability to implement the model. Issues that must be considered include the difficulty of obtaining the model, the level of documentation and testing associated with the model, and the ease of model use. The budget and the schedule for any project will affect the type of criterion used and ultimately the model selected (Environmental Protection Agency, 1988c).

The 1988 guidance represents a significant advance in the EPA modeling program because it provides structure to the model selection process and will avoid mixing discussions of model applicability and model results. Dividing these two processes could help simplify interactions between the regulators and the regulated community. Even this guidance represents only a small step toward simplifying the regulatory process. A number of codes used in EPA programs are described in the latter portion of the report. However, information on the level of complexity of these codes and the criteria for their application are not included. Additional clarification of EPA model use will be needed to help direct code selection in model applications that will be submitted to the agency.

If problems arise, EPA personnel are directed to EPA's Center of Exposure Assessment Modeling and the International Groundwater Modeling Center for specific advice (Environmental Protection Agency, 1988b, p. 3-33; 1988c). Ultimately, however, EPA personnel must rely upon their own skills.

Resource Conservation and Recovery Act

Law and Regulations

There are tens of thousands of facilities that handle hazardous waste and therefore must obtain a permit. The Resource Compensation and Recovery Act (RCRA) establishes comprehensive, "cradle-to-grave" hazardous waste management programs. RCRA forbids waste treatment or disposal and limits waste storage for facilities not holding appropriate permits from EPA or a state agency (Section 3005[a] of RCRA, 42 USC §6925[a]).

The very foundation of any regulatory program is the definition of what is regulated versus what is not. The EPA definition of a hazardous waste determines "whether a waste, if mismanaged, has the potential to pose a significant hazard to human health or the environment due to its propensity to leach toxic compounds" (51 Fed. Reg. 21,653 [1986][6]). EPA has listed industrial waste streams as hazardous based on a limited sampling of a representative number of plants in the industry. Also, a waste is considered hazardous if it is ignitable, corrosive, or explosive or if the leachable concentrations of certain chemicals exceed regulatory health-based limits (i.e., the extraction procedure [EP] test). A waste is hazardous based on this EP test if chemicals will leach out of the waste in quantities that may cause the ground water concentrations 500 ft downgradient to exceed drinking water standards after the waste is placed in a municipal landfill. EPA's original definition of hazardous waste assumed arbitrarily that the leachable concentration of a chemical would decrease by a factor of 100 in the 500 ft (45 Fed. Reg. 33,084 [1980][7]).

In 1986, EPA proposed to modify the EP test used to define hazardous waste by, among other things, (1) adding 38 organic chemical constituents, (2) substituting a more rigorous leaching test, (3) applying compound-specific attenuation and dilution factors for each organic constituent to evaluate the worst-case potential impact on ground water 500 ft downgradient of the location of disposal, and (4) using a risk-based concentration when no drinking water standard is available (51 Fed. Reg. 41,082 [1986][8]). EPA's proposed new definition uses a subsurface fate and transport model, called EPASMOD (or the Composite Landfill Model), to derive compound-specific attenuation and dilution factors. EPASMOD considers the dilution, hydrolysis, and soil adsorption that occur as a chemical migrates from the bottom of a landfill to a drinking water source 500 ft away (see 51 Fed. Reg. 1,602 [1986][9] for a more detailed discussion of the EPASMOD).

The Environmental Protection Agency has revised EPASMOD and its input data and is considering additional revisions to EPASMOD and its input data so that the predicted concentrations would be less overpredictive (53 Fed. Reg. 28,892 [1988][10]). The proposal therefore would incorporate a contaminant transport into the definition of hazardous waste.

Contaminant transport models also have been used in other aspects of the RCRA program. For example, EPA uses the vertical-horizontal spread (VHS) model to determine when a listed hazardous waste from a particular facility would no longer be subject to RCRA hazardous waste requirements because the particular characteristics of the waste from that facility make the waste nonhazardous wherever it may be disposed (50 Fed. Reg. 48,886 [1985]; see Case Study 1, below). The RCRA regulations also require the permittee to perform ground water monitoring (40 CFR §264.97, 264.98, 264.99) and to clean up contaminated conditions at active facilities in any area where there was historic disposal of either hazardous or solid wastes (40 CFR §264.100). The corrective action requirements, in essence, convert RCRA into a Superfund-type cleanup statute and expand RCRA's jurisdiction to cover all inactive waste disposal areas on operating facilities.

The RCRA regulations require a permittee to clean up the ground water to (1) background levels, (2) the concentrations specified by EPA for drinking water, or (3) a site-specific risk-based action level (the alternate concentration limit, or ACL) (40 CFR §264.94). To evaluate the potential adverse effects on ground water quality, the permittee must provide information on, among other things, the wastes' potential for migration; the hydrogeological characteristics of the facility and surrounding land; the existing quality of ground water, including other sources of contamination and their cumulative impact on the ground water quality; and the potential for health risks caused by human exposure to waste constituents (40 CFR §264.94[b]).

The permittee must also submit an exposure and risk assessment. The two key concepts in the ACL process are that (1) the cleanup level must protect the public at the point of exposure (i.e., where ground water is withdrawn to use as drinking water), and (2) the point of compliance (i.e., the point where ground water is monitored) must be at the boundary of the regulated unit (Environmental Protection Agency, 1987a).

It is necessary to set the ACL at a level (usually monitored at the boundary of the regulated unit) that, based on predictions, will result in ground water exposures that are below health protective levels at some distant point (e.g., the nearest drinking water well) and some future time.

Guidance

Contaminant transport modeling can be used for the same purposes as in the Superfund program. EPA's general exposure guidance concerning the use of models (above) is equally applicable here. The RCRA guidance encourages using conservative assumptions "where time and/or resources are limited" (Environmental Protection Agency, 1986b, p. 150). Numerical models are preferred over analytical models (p. 158). EPA's RCRA guidance lists publicly available models (p. 158).

The Environmental Protection Agency's RCRA guidance is contradictory, however. EPA's RCRA alternative concentration limit guidance (Environmental Protection Agency, 1987a, p. 4-6) states that

> [a]lthough not required for an ACL demonstration, mathematical simulation models of ground water flow and contaminant transport can be extremely useful tools for the applicant. Models are more appropriate for relatively simple geologic environments where conditions do not vary widely; in complex geologic areas, modeling may be less useful.
>
> The permit applicant is responsible for ensuring that the models used simulate as precisely as possible the characteristics of the site and the contaminants and minimize the estimates and assumptions required. . . . *Whenever possible, input parameters and assumptions should be conservative in nature; worst-case scenarios may save much effort.* [Emphasis added.]

The RCRA ground water monitoring guidance, on the other hand, states "modeling results should *not* be unduly relied upon in guiding the placement of assessment monitoring wells or in designing corrective actions" (Environmental Protection Agency, 1986b, p. 156; emphasis added).

Recently, EPA has considered standardizing the steps in the risk assessment/modeling process by "prescribing the types of models that can be used or the assumptions that are incorporated into models" (Environmental Protection Agency, 1987b). Among the standard models being considered is the VHS model. The standard model would guide the decision "based on only minimal site-specific data" (Environmental Protection Agency, 1987b). The use of a nationwide database would be contrary to EPA site-specific use on the selection of models.

Underground Injection Control (UIC) Program

Law and Regulations

The Safe Drinking Water Act prohibits underground injection unless such injection is authorized by a permit or by rule (Section 1421, 42 USC §300h). The Underground Injection Control (UIC) regulations govern, among other wells, Class I wells, those wells used to dispose of hazardous waste below an underground source of drinking water. Class I wells are subject to regulations that specify minimum design construction and operating conditions and require continued monitoring of the nearby ground water to ensure that a present or future drinking water supply is not endangered (40 CFR Part 144).

The Environmental Protection Agency's recent amendments to the Class I well regulations prohibit the injection of hazardous waste into wells unless (1) the waste is treated to the same extent required for hazardous waste disposed of on land, or (2) EPA grants an exemption from the regulation based on a "no migration" petition (53 Fed. Reg. 28,118 [1988][11]). The burden is on the permit applicant to prove that no migration will occur.

A petitioner must demonstrate that migration outside the injection zone will not occur for 10,000 yr (53 Fed. Reg. 28,155). Nothing in the statute or its legislative history forbids the use of models or requires their use (53 Fed. Reg. 28,126). These regulations *require* the person seeking a "no migration" exemption to submit "predictive models" that are "appropriate for the specific site, waste streams, and injection conditions of the operation, and shall be calibrated for existing sites where sufficient data are available" (53 Fed. Reg. 28,156). The petitioner also must use "reasonably conservative values," an approved "quality assurance and quality control plan," and a "sensitivity analysis to determine the effect that significant uncertainty may contribute to the demonstration" (53 Fed. Reg. 28,156).

The Environmental Protection Agency rejected the contention that one could not accurately model over a 10,000-yr period (53 Fed. Reg. 28,126). In determining the feasibility of a 10,000-yr goal, EPA concluded that (1) the "modeling need not locate the exact point where the waste would be . . . [in 10,000 yr]; determining where it would not be [i.e., outside the injection zone] is sufficient. This level of precision is achievable" (53 Fed. Reg. 28,126[12]); (2) such fluid flow modeling was considered "a well-developed and mature science" that had been "used for many years in the petroleum industry" (53 Fed.

Reg. 28,127); (3) such models had been developed by the DOE for use in the nuclear waste isolation program; and (4) this model and its application were peer reviewed by EPA's Science Advisory Board (a group of independent scientists who advise EPA in scientific issues). The Science Advisory Board (1984, cited in 52 Fed. Reg. 32,446) concluded that

> [m]odeling for the time periods involved . . . required extension of such . . . techniques well beyond usual extrapolation, however, the extension for 10,000 years can be made with reasonable confidence.

From EPA's policy point of view, the precision of these predictions is not the only issue. The intent of the statute is to allow deep well injection of chemicals as long as they will not migrate outside the injection zone for a very long period of time. The use of models might be considered a failure if chemicals actually migrate outside the injection zone in 1 or even 30 yr. If a petition were granted, but chemicals migrated outside the injection zone in 8,000 yr instead of 10,000 yr, the overall purpose of the regulation would still be served. The model might be considered by many to be satisfactory.

As a practical matter in this situation, the only choices other than using contaminant transport models would be to rely on the best professional judgment of a qualified hydrologist or provide for no exemptions to the ban on hazardous waste disposal. Because the statute provides for such exemptions, EPA has attempted to balance the scientific uncertainty.

Guidance

As of the time this report was written, this program had not yet developed guidance for this use of models.

Conclusion

The Environmental Protection Agency's guidance is contradictory. Some guidance provides a rational scientific framework for selecting models, and other guidance appears to favor use of standardized worst-case models.

SELECTED CASE STUDIES

The five case studies in this section are presented as examples of how contaminant transport models are currently used as tools to (1) understand ground water systems, (2) predict contaminant migration, and (3) illustrate how models are used by regulatory agencies in the decisionmaking process. The case studies were chosen to involve a large number of the hydrogeologic processes discussed in Chapters 3 and 4, and to demonstrate how this knowledge of hydrogeologic processes is used in actual problem-solving applications. A common theme throughout the case studies is that there is a lack of knowledge about system parameters. The case studies illustrate several methods that may be used to deal with this uncertainty.

The selection of case studies is inherently subjective. The committee decided that the selection process could not totally exclude controversial examples. By their nature, many of the best-documented uses of models involve problems where there is a factual dispute that is longstanding or involves significant issues, e.g., millions of dollars in remedial costs or the right to use a scarce resource such as water.

Selection of a particular case study should not be misconstrued as a judgment by the committee concerning whether the particular model was appropriately selected or applied. It is not possible to make such determinations without an extensive evaluation of the facts. Such a case-specific, detailed evaluation is beyond the scope of this report and not necessary to accomplish the committee's task. Therefore nothing in the report should be construed as a definitive scientific evaluation or endorsement of any particular model or modeling approach.

The case studies cover a wide range of ground water problems (Table 5.1), involving sites scattered across the United States (Figure 5.1). The first case study, the VHS model, discusses the use by EPA of a generic model (i.e., a model that does not require site-specific information) to determine which solid wastes should be treated as hazardous wastes.

The Madison aquifer case illustrates the use of a variety of ground water flow modeling approaches to predict water-level declines from large well withdrawals in an aquifer system in which very little is known about the hydraulic properties of the aquifer (Konikow, 1976). Accurately predicting ground water flow conditions is an essential first step in simulating contaminant transport in ground water, and this case study illustrates particularly well techniques that can be used to assess the reliability of predicted ground water flow conditions.

TABLE 5.1 Synopsis of Case Studies

Short Title	Subject
VHS model	A generic ground water transport model used by EPA to predict contaminant migration. The use of this model to delist wastes from the Gould, Inc., facility in McConnelsville, Ohio, is discussed.
Madison aquifer	Evaluation of large ground water withdrawals from an aquifer using a ground water flow model when aquifer parameters are poorly known.
Snake River plain	A ground water transport model used to predict migration of chloride, tritium, and strontium-90 in basalts. Original modeling study conducted in 1973 predicted concentrations in years 1980 and 2000. Subsequent study conducted in 1980 evaluated accuracy of original predictions.
Tucson Airport	The use of a ground water flow and transport model to assign liabilities for a multisource plume to specific sources.
S-Area	A one-dimensional, two-phase flow model used to evaluate the migration of nonaqueous-phase liquids at the S-Area landfill, Niagara Falls, New York.

The Snake River plain case study discusses a simulation of chloride migration in ground water conducted in 1974 using a numerical ground water transport model and a subsequent field study conducted in 1980 to check the model predictions (Lewis and Goldstein, 1982). This case study is one of a few ground water contamination problems in which field data have been collected almost a decade

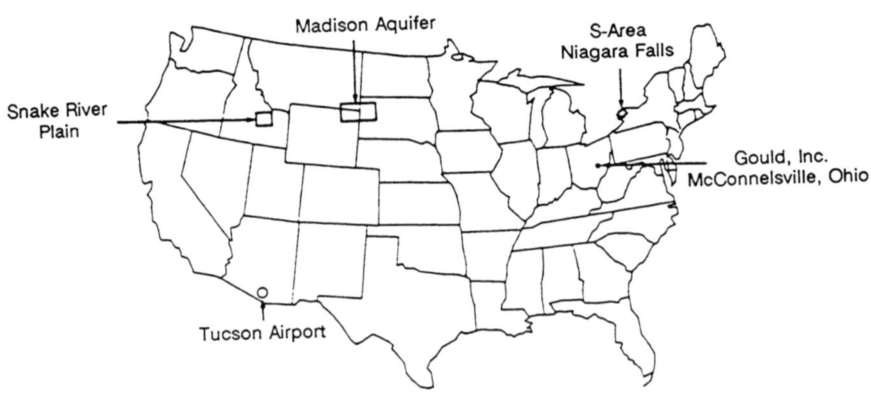

FIGURE 5.1 Location of case studies.

after the modeling was completed to compare the predictions to the observed concentrations. This study illustrates the error that can be expected with predictions of contaminant migration and the problems that result when a three-dimensional ground water system is modeled as a two-dimensional system.

The Tucson Airport study discusses how a ground water model was used to assign liabilities to individual parties for specific ground water contamination incidents.[13] Models are frequently being used for this purpose at Superfund sites (sites on the National Priority List) at which there are several responsible parties. The discussion that follows this case study highlights the conceptual problems that the committee foresees as a result of using the current generation of ground water transport models for this purpose.

The S-Area case study examines the use of models to investigate the migration of a immiscible, denser-than-water fluid within an aquifer. The models discussed in this study were developed to help design an appropriate remedial action for the site. The models were presented in litigation involving this site and have been explicitly incorporated into a legally enforceable document as the method to be used in designing the remedial action for the site.

To ensure consistent emphasis of particular points, a general format was adopted for the presentation of the case studies. The preparer of each case study was asked to treat in sequence, if possible, the objective of the study, the major hydrogeologic processes considered in the study, a brief description of the model used, and the results and conclusions of the study. Each case study is followed by a committee discussion on the strengths and weaknesses of the study and the lessons that can be drawn from it for other studies of a similar nature.

Vertical-Horizontal Spread (VHS) Model Background

As described above, EPA regulations allow a particular plant within the industry to demonstrate that its particular wastes are nonhazardous, i.e., to delist the particular wastes (40 CFR §§260.20, 260.22). The primary quantitative criterion used to evaluate delisting petitions is whether, assuming worst-case conditions, the leachable chemicals from the solid waste would result in unacceptable ground water quality 500 ft downgradient of the disposal location. As a matter of policy, EPA uses the VHS model (Domenico and Palciauskas, 1982) to determine the concentration of the leachable chemicals 500 ft downgradient from the location of disposal for delisting purposes. The toxicity of the waste is evaluated by comparing the concentrations estimated by the VHS model for a location 500 ft downgradient with EPA drinking water standards or other health-based standards (51 Fed. Reg. 21,666; also 53 Fed. Reg. 18,025 [1988][14]). If the concentration at the well is lower than the standard, the waste is considered nonhazardous and will be delisted.

Model

The model used in the delisting process considers three basic steps (Domenico and Palciauskas, 1982; 50 Fed. Reg. 48,886 [1985]; 50 Fed. Reg. 7882 [1985][15]; 50 Fed. Reg. 41,082 [1986][16]):

- generation of a leachate from the waste;
- migration of the leachate to an underlying ground water aquifer; and
- migration of the contaminated ground water in the aquifer to a nearby drinking water well.

The concentrations of the chemical compounds of interest are generated by an appropriate leaching test, e.g., the extraction procedure test (40 CFR §§261.24) or a leaching estimation method such as the organic leachate model (OLM) (51 Fed. Reg. 21,653 [1986]). The extraction procedure (EP) and the toxicity characteristic leaching procedure (TCLP) are laboratory procedures that are designed to simulate codisposal of the waste with municipal wastes in a sanitary landfill. The OLM is an empirical equation that calculates leachate concentration of a compound on the basis of the compound's solubility and the concentration of the compound in the waste.

The second step of the model estimates the attenuation that may occur during the migration of the leachate from the waste to the underlying aquifer. EPA assumes that no attenuation occurs because this is a reasonable worst-case characteristic of saturated soil systems, and because the water table is near the bottom of many waste sites.

The third step of the modeling process calculates the dispersion of the chemical compound in a drinking water aquifer in the vertical and horizontal directions perpendicular to ground water flow as a result of a continuous source of contamination. The VHS model is used to simulate the dispersion of the contaminants and calculate the contaminant concentration at a reception well directly downgradient

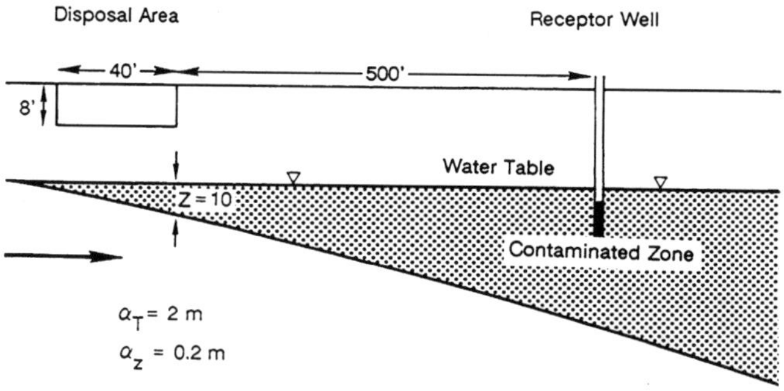

FIGURE 5.2 Schematic of vertical-horizontal spread model.

of the waste disposal area (Figure 5.2). The following equation describes the VHS model:

$$C_y = C_o \mathrm{erf}(Z/[2(\alpha_z Y)^{0.5}])\mathrm{erf}(X/[4(\alpha_T Y)^{0.5}]),$$

where

C_y = contaminant concentration at the receptor well (mg/l)
C_o = contaminant concentration in the leachate (mg/l)
erf = error function, dimensionless
Z = penetration depth of leachate into the aquifer (m)
Y = distance from disposal site to compliance point (m)
X = length of the disposal site measured in the direction perpendicular to the direction of ground water flow (m)
α_T = lateral transverse (horizontal) dispersion length (m)
α_z = vertical dispersion length (m)

This equation has three basic terms: (1) the initial concentration of the contaminant in the leachate, (2) a term for the spreading of the concentration in the vertical direction, and (3) a term for the spreading of the contaminant in the horizontal dimension. Vertical and horizontal spreading are the only processes that cause the contaminant concentration to decrease away from the source. Other processes, such as chemical reactions, precipitation, and biodegradation, that might decrease contaminant concentrations as they migrate away from the source are not considered in the VHS model.

The VHS model is a steady-state model, and the calculated receptor well concentration is a steady-state concentration. The model does not calculate the time required to reach the steady-state concentrations. If the contaminant is strongly sorbed to aquifer materials, such as is the case for polychlorinated biphenyls (PCBs) and dioxins, the time required to reach steady state will be thousands of years.

Use of the VHS model requires the specification of only six parameters: initial leachate concentration (C_o), distance to receptor well (Y), length of disposal site perpendicular to direction of ground water flow (X), horizontal transverse dispersion length (α_T), vertical dispersion length (α_z), and the mixing zone depth (Z). EPA uses reasonable worst-case values for these parameters, except for leachate concentration and the length of the disposal site, which vary with each delisting petition.

The reasonable worst-case parameter values used by EPA are as follows:

- *Distance to well*: EPA uses 500 ft for the distance from the waste disposal facility to the drinking water well. This distance is based on an informal survey that suggested that at 75 percent of the landfills, the closest well is further than 500 ft (152.4 m) from the facility.
- *Dispersion lengths*: A value of 6.5 ft (2 m) is specified for the horizontal transverse dispersion length, and 0.65 ft (0.2 m) is specified for the vertical dispersion length.
- *Mixing zone depth*: A mixing zone depth of 10 ft (3.28 m) is used by EPA. The mixing zone depth is related to the width of the disposal area and the ratio of leachate generation to the velocity of ground water. EPA assumed that the average disposal area width was 40 ft (12.2 m) and that the ratio was 0.25.
- *Length of disposal area*: The length of the disposal area is calculated by using the waste volume specified in the delisting petition and assuming that the waste is placed in a trench 40 ft (12.2 m) wide and 8 ft (2.4 m) deep, where the long axis is oriented perpendicular to the direction of ground water flow. A minimum of length of 40 ft is used if waste volume is small.

When the reasonable worst-case parameter values are substituted into the VHS equation, a relatively simple equation relating a dilution factor (C_o/C_y) to the waste volume results. A graph of the dilution factor versus waste volume is shown in Figure 5.3. This graph shows that a solution factor of 32 is calculated from the VHS model for small waste volumes and that the calculated dilution factor decreases to about 7 for large waste volumes.

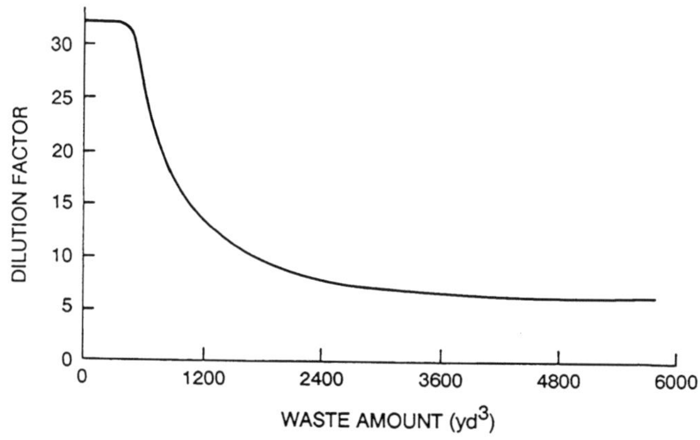

FIGURE 5.3 **Predicted dilution factor as a function of waste amount.**

Application at a Particular Site

Gould, Inc. operates an electroplating facility in McConnelsville, Ohio, that annually generates 1,100 yd^3 of sludge from its wastewater treatment plant. This sludge is classified as a hazardous waste in 40 CFR §260.22, and Gould, Inc. petitioned to delist the waste based on the destruction and immobilization of hazardous compounds by its wastewater treatment system.

The constituents of concern in the sludge are cadmium, chromium, and nickel. Gould, Inc. submitted leaching test data for these compounds, and the VHS model was used to determine concentrations at the receptor well using a dilution factor of 13.5, which was calculated from the waste volume, and the concentrations of these compounds were set to health-based standards (see Table 5.2).

The calculated concentrations of cadmium, chromium, and nickel at the receptor well are all below the health-based standard. EPA used this evaluation to delist this waste (50 Fed. Reg. 48,887 [1985]).

TABLE 5.2 RCRA Delisting Data for Gould, Inc., Facility, McConnelsville, Ohio

Chemical Compound	Leachate Concentration (mg/l)	Calculated Receptor Well Concentration (mg/l)	Health-Based Standard (mg/l)
Cadmium	<0.1	<0.007	0.01
Chromium	<0.5	<0.037	0.05
Nickel	2.5	0.185	0.35

Regulatory Context

The Environmental Protection Agency adopted the use of the VHS model in the RCRA delisting program as a matter of policy,[17] but claimed that it was not bound to use the model (50 Fed. Reg. 7,882 (1985)). EPA has "treated the model as conclusively disposing of certain issues. . . . The model thus created a norm with 'present-day binding effect' on the rights of . . . [the companies seeking to have a waste delisted]" (838 F.2d at 1321). In practice, EPA "evidenced almost no readiness to reexamine the basic propositions that make up the VHS model . . . " (838 F.2d at 1321). As a result, a unanimous panel of the Court of Appeals ruled that EPA violated the Administrative Procedure Act (5 USC 7=1 706). EPA is now obligated, in reality, to exercise discretion in individual delisting cases or to issue the VHS as a binding regulation after notice and public comment and an opportunity to challenge the regulation (838 F.2d at 1324). EPA, however, continues to use the VHS model in the RCRA delisting program purportedly now as a truly nonbinding policy (53 Fed. Reg. 21,640 [1988][18]).

Discussion

The VHS model is a simple generic model. The use of the model requires no site-specific data, and therefore it may appear unscientific. EPA (53 Fed. Reg. 7,906 [1988][19]) openly acknowledges that

> the VHS model is more likely to overpredict (rather than underpredict) the receptor concentration of contaminants in any given waste due to the conservative nature of the assumptions underlying the model. EPA also recognizes that all models do not always predict factual values accurately.

The Environmental Protection Agency's position (53 Fed. Reg. 7,906 [1988]) is that

> [u]nless the Agency is able to assure protection of human health and the environment without generic, conservative assumptions, the Agency will employ these assumptions.

The Environmental Protection Agency is particularly concerned that once a waste is delisted, there are no restrictions on how or where the waste will be disposed. The model, however, could be improved by the addition of chemical-specific terms for biodegradation, precipitation, and other reactions that would cause the concentration of the contaminant to decrease as it migrates from the source area to the receptor well. Health protective values could be used instead of being ignored altogether.

As described above, EPA has proposed using the EPASMOD to define hazardous wastes. Although EPA rejected the use of EPASMOD for delisting petitions in 1986, it also indicated that it might reconsider use of that model once the new test for hazardous wastes was completed (51 Fed. Reg. 41,501 [1986][20]). More recently, EPA has considered the possible use of EPASMOD in the delisting process (53 Fed. Reg. 28,892 [1988][21]). This model would take into account at least some of the factors that decrease concentration in the real world. EPA would still not use site-specific data in the model. This case study illustrates how easily a nonbinding guidance can in reality become an inflexible rule.

Madison Aquifer—Well Withdrawals from a Deep Regional Aquifer

Background

The Powder River basin of northeastern Wyoming and southeastern Montana contains large coal reserves that have not yet been fully developed. The future development of these energy resources will be accompanied by increased demands for water, which is not abundantly available in this semiarid area. One plan had been formulated to construct a coal slurry pipeline to transport coal out of the area; it would have required approximately 15,000 to 20,000 acre-ft/yr of water (20 to 28 ft^3/s). In the mid-1970s a plan was proposed to supply water for the coal slurry pipeline by withdrawals from up to 40 deep wells that would be drilled about 3,000 ft into the Mississippian Madison limestone in Niobrara County, Wyoming. The Madison aquifer is an areally extensive Paleozoic carbonate rock system that underlies an area exceeding 100,000 mi^2 in the Northern Great Plains. Wyoming authorized the withdrawals, but the state of South Dakota was concerned about the cross-boundary effects of the drawdown.

Initial Modeling Studies

Large ground water withdrawals may cause significant water-level declines in the Madison aquifer, perhaps extending into adjacent states, as well as decreases in streamflow and spring discharge in or near the outcrop areas. Thus an ability to predict the effects of the proposed ground water withdrawals on potentiometric levels, recharge, and discharge is needed.

The Madison aquifer lies at great depths (between 1,000 and 15,000 ft) in most of the area and is therefore relatively undeveloped. There are insufficient data available to accurately and precisely define the head distribution and the hydraulic properties of the aquifer. In light of this uncertainty, and as a prelude to a planned subsequent 5-yr hydrogeologic investigation of the Madison aquifer, Konikow (1976) developed a preliminary digital model of the aquifer using the two-dimensional finite-difference model of Trescott et al. (1976). The objectives of the preliminary model study were to (1) improve the conceptual model of ground water flow in the aquifer system; (2) determine deficiencies in existing data and help set priorities for future data collection by identifying the most sensitive parameters, assuming the model is accepted as being appropriate; and (3) make a preliminary estimate of the regional hydrologic impacts of the proposed well field.

Initial Results

The results indicated that the aquifer can probably sustain increased ground water withdrawals up to several tens of cubic feet per second, but that these withdrawals probably would significantly lower the potentiometric surface in the Madison aquifer in a large part of the basin. The model study and predictions were framed in terms of a sensitivity analysis because of the great uncertainty in most of the parameters. For example, Figure 5.4 shows drawdown predictions made for an area near the proposed well field for an assumed reasonable range of values for the storage coefficient (S) and leakage coefficient ($K_z m$), where K_z and m are the vertical hydraulic conductivity and the thickness, respectively, of the confining layer. The curves show that the range in plausible drawdowns, even after 1 yr, is extremely large.

This uncertainty in the nature and magnitude of potential vertical leakage was also translated into a disparity of interpretations and opinions in other independent forecasts of these impacts. In the report of the 1975 hearings before the U.S. Congress on pending legislation pertaining to the coal slurry pipeline, a report by the consultants to the pipeline company concludes that ". . . the 'leakage' or contribution from beds adjacent to the porous zones in the Madison is sufficient to preclude drawdown at distances more than about 2000

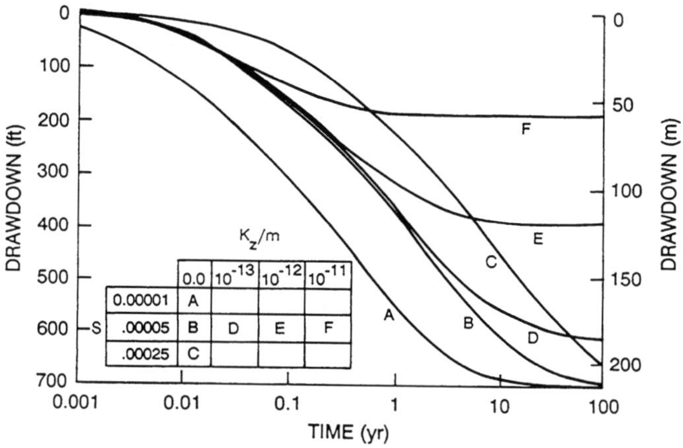

FIGURE 5.4 Time-drawdown curves for model node near hypothetical pumping wells in the Madison limestone (modified from Konikow, 1976).

feet." On the other hand, the same report of the 1975 congressional hearings includes a disparate forecast by another expert, based on assumption of nonleaky conditions, which shows drawdowns after 45 yr exceeding 1,000 ft near the pumping center and greater than 200 ft at distances, more than 50 mi away.

Other Model Studies

This preliminary model analysis helped in formulating an improved conceptual model of the Madison aquifer. For example, the important influences of temperature differences and aquifer discontinuities on ground water flow in the Madison were recognized and documented as a result of the model analysis. It could be argued that the importance of these influences could have been (or should have been) recognized on the basis of hydrogeologic principles without the use of a simulation model. However, none of the earlier published studies of this aquifer system indicated that these factors were of major significance. The difference from earlier studies arose from the quantitative hypothesis-testing role of the model; the nature of the inconsistencies between observed head distributions and those calculated using the initial estimates of model parameters helped direct the investigators toward testing hypotheses that would resolve or minimize the inconsistencies. Also in this case, the demonstrated high sensitivity to the leakage coefficient highlighted the need to

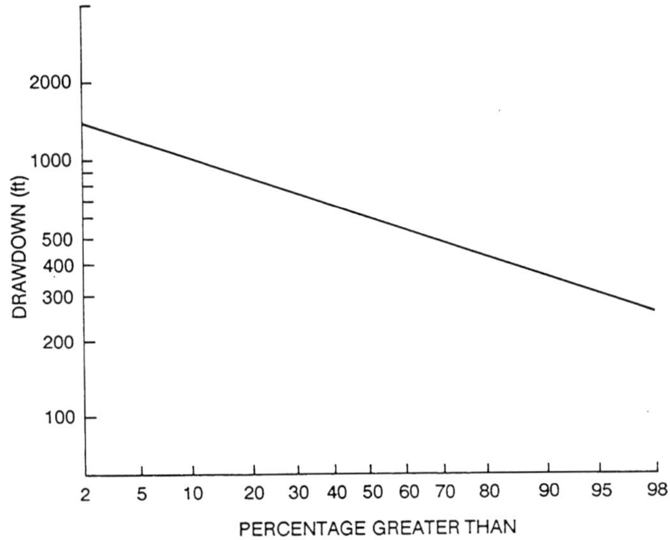

FIGURE 5.5 Calculated probability distribution of drawdowns at the Niobrara well field.

reevaluate the system in a true three-dimensional framework so as to better consider vertical components of flow.

The effects of the pumping for the coal slurry pipeline were reexamined by Downey and Weiss (1980), and Woodward-Clyde Consultants (1981), with a three-dimensional model that incorporated the processes found to be important in the initial study. The latter study, which was prepared for an environmental impact statement for the project, used a five-layer model and a Monte Carlo simulation approach to incorporate and assess the effects of uncertainties in the parameters. The predicted impacts were then presented as probability distribution curves showing the likelihood of different drawdowns occurring at the specified points (Figure 5.5). The recognized uncertainty in the predictions (i.e., the wide range in predicted drawdowns) was a factor contributing to the fact that the coal slurry pipeline was never built.[22] The controversy surrounding the effects of the proposed drawdown in South Dakota was a major factor in the pipeline company's decision to buy Missouri River water from the state of South Dakota to supply the pipeline. Because of falling coal prices and railroad opposition, the project was abandoned in 1984.

Regulatory Context

The original models developed of the Madison aquifer were used in testimony presented at hearings before a congressional subcommittee to support the viewpoints of the proponents and the opponents of the coal slurry pipeline. Because field data on aquifer properties were sparse, a wide range of parameter values was probable. The opponents of the project chose parameter values that resulted in the prediction of widespread impacts, and the proponents chose parameter values that resulted in the prediction of minimal impacts. The hydrogeologic parameters used in both models were not inconsistent with the available data, but parameter values were clearly chosen to bias the predictions. These modeling approaches illustrate the disparity of results that can be predicted using models when a rigorous approach is not utilized to analyze parameter uncertainty.

The later model developed for the environmental impact statement, which was prepared for the Bureau of Land Management, was prepared under fairly rigid guidelines that were designed to produce an objective analysis of the potential environmental impacts of the proposed project. This modeling study is clearly a more objective analysis of the probable impacts of the proposed ground water withdrawals.

Discussion

Model analyses and predictions can lead to an improved understanding of an aquifer system and serve as an aid to making decisions or formulating policy. However, the predictions must be clearly presented, together with a realistic assessment of the confidence in them. This case study demonstrates that models can be very useful tools for gaining an understanding of aquifer systems in which little is known about the hydraulic properties of the system and that these models can be invaluable for prioritizing field data collection activities so that a maximum amount of information can be obtained for a given expenditure.

Accurately predicting ground water flow conditions is an essential first step in simulating contaminant transport in ground water, and this case study illustrates, particularly well, techniques that can be used to assess the reliability of predicted ground water flow conditions. Because most of the ground water contamination problems of interest to regulators are dominated by convective transport, the uncertainty associated with predictions of ground water transport at these sites can be assessed using techniques similar to those used in this case study.

Snake River Plain—Point Source of Contamination

Background

The Idaho National Engineering Laboratory (INEL) is located on 890 mi^2 of semiarid land in the eastern Snake River plain of southeast Idaho. The facility, formerly called the National Reactor Testing Station, is now operated by the Department of Energy for testing various types of nuclear reactors. Robertson (1974) reports that several facilities at the site generate and discharge low-level radioactive and dilute chemical liquid wastes to the subsurface through seepage ponds and disposal wells. The two most significant waste discharge facilities, the Test Reactor Area (TRA) and the Idaho Chemical Processing Plant (ICPP), have discharged wastes continuously since 1952. The purpose of this study was to predict the future migration of wastes containing chloride, tritium, and strontium-90. Unlike the other case studies discussed in this section, the modeling analyses of the INEL were not conducted to satisfy the requirements of a regulatory agency. This discussion of the INEL site and associated model is largely extracted and paraphrased from the reports of Robertson (1974) and Lewis and Goldstein (1982), to which the reader is referred for additional details.

Hydrogeologic Setting

The eastern Snake River plain is a large structural and topographic basin about 200 mi long and 50 to 70 mi wide. It is underlain by 2,000 to 10,000 ft of thin basaltic lava flows, rhyolite deposits, and interpolated alluvial and lacustrine sediments. These formations contain a vast amount of ground water and make up the major aquifer in Idaho, which is known as the Snake River plain aquifer. Ground water flow is generally to the southwest at relatively high velocities (5 to 20 ft/day). The principal water bearing zones occur in the basalts, the permeability fabric of which is highly heterogeneous, anisotropic, and complicated by secondary permeability features, such as fractures, cavities, and lava tubes.

Model Formulation

Concern over ground water contamination from the waste discharge prompted Robertson (1974) to develop a digital solute transport model to simulate the underlying aquifer system. The model (that is, the numerical method used to solve the solute-transport equation) was based on the method of characteristics. Robertson first calibrated a flow model for a 2,600-mi^2 area and then calibrated the transport model for a smaller part of that area in which contamination was of concern. The calibration of the transport model was based on a 20-yr history of contamination, documented by samples from approximately 45 wells near and downgradient from the known point sources of contamination. These data showed that chloride and tritium had spread over a 15-mi^2 area and migrated as far as 5 mi downgradient from discharge points. The distribution of waste chloride observed in November 1972 is shown in Figure 5.6. Robertson notes that the degree of observed lateral dispersion in the plumes is particularly large.

Results and Conclusion

Robertson used the calibrated transport model to predict future concentrations of chloride, tritium, and strontium-90 for the years 1980 and 2000 under a variety of possible future stresses. For the chlorides, assumptions included were that (1) disposal continues at 1973 rates and (2) the Big Lost River recharges the aquifer in odd-numbered years. This scenario came closest to what actually occurred. The projections indicated that by 1980 the leading edges of both the chloride (see Figure 5.7) and the tritium plumes would be at or near the INEL boundary.

Lewis and Goldstein (1982) report that eight wells were drilled during the summer of 1980 near the southern boundary to help fill data gaps and to monitor contaminants in ground water flowing across the INEL boundary. They also used the data from the eight wells to help evaluate the accuracy of Robertson's predictive model. The distribution of waste chloride observed in October 1980 is shown in Figure 5.8. A comparison of Figure 5.8 with Figure 5.6 indicates that the leading edge of the chloride plume had advanced 2.5 to 3 mi during that 8-yr period and that the highest concentrations increased from 85 mg/l to around 100 mg/l.

A comparison of Figures 5.7 and 5.8 indicates that although the observed and predicted plumes show general agreement in the

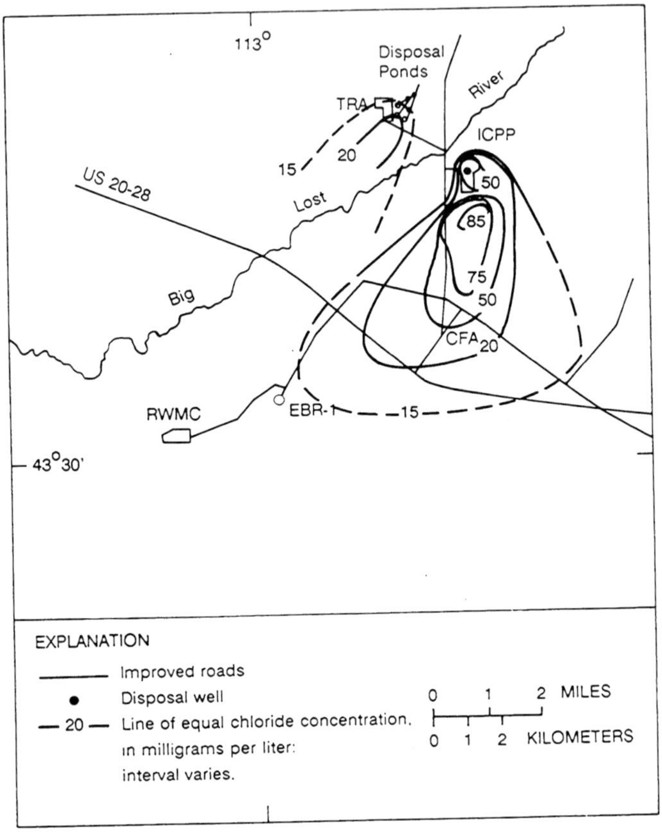

FIGURE 5.6 Observed distribution of waste chloride in ground water in the Snake River plain aquifer, Idaho, ICPP-TRA vicinity in 1972. SOURCE: Robertson, 1974.

direction, extent, and magnitude of contamination, some apparently significant differences in detail exist. The observed plume is broader and exhibits more lateral spreading than was predicted and has not spread as far south and as close to the INEL boundary as predicted. Also, the predicted secondary plume north of the Big Lost River, emanating from the Test Reactor Area, was essentially not detected in the field at that time.

Lewis and Goldstein (1982) presented a number of factors that they believed contributed to the discrepancy between predicted and observed results. These reasons can be summarized as follows: (1) there was less dilution from recharge during 1977 to 1980 because

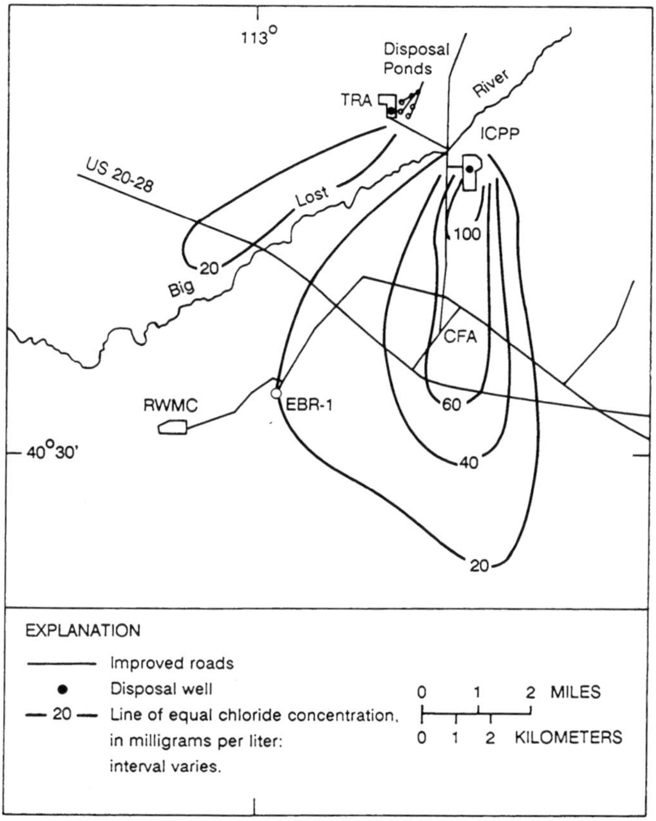

FIGURE 5.7 Model-projected distribution of waste chloride in the Snake River plain aquifer for 1980 (ICPP-TRA vicinity), assuming disposal continues at 1973 rates and the Big Lost River recharges the aquifer in odd-numbered years. SOURCE: Robertson, 1974.

of below-normal river flow; (2) chloride disposal rates at the ICPP facility were increased during the several years preceding 1980; (3) the model grid may have been too coarse; (4) the model calibration selected inaccurate hydraulic and transport parameters; (5) vertical components of the flow and transport may be significant in the aquifer but cannot be evaluated with the two-dimensional areal model; (6) there may be too few wells to accurately map the actual plumes, and some existing wells may not be constructed properly to yield representative measurements; and (7) the numerical method introduces some errors (however, Grove's [1977] analysis of this same

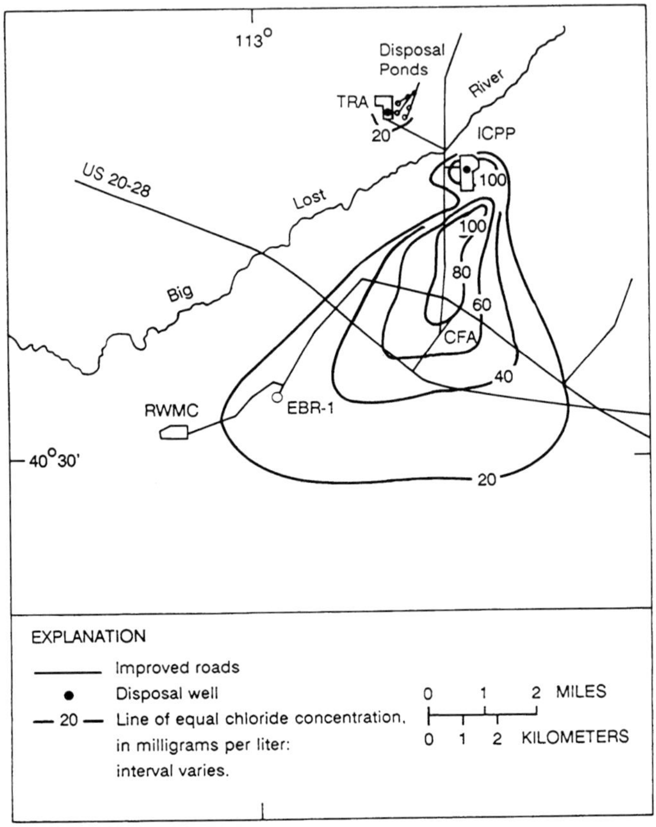

FIGURE 5.8 Distribution of waste chloride in the Snake River plain aquifer (ICPP-TRA vicinity), October 1980. SOURCE: Lewis and Goldstein, 1982.

system used finite-difference and finite-element methods, and comparisons of numerical results offer no basis for concluding that the numerical solution algorithm used by Robertson was in itself a significant source of the predictive errors). Although these factors can be expanded upon, and other factors added, it is extremely difficult to assess the contribution of any single factor to the total error. Recalibration of the earlier model using the now extended historical record could be employed to test some of these hypotheses. Other factors can only be tested if new models are developed that incorporate additional or more complex concepts, such as density differences and three-dimensional flow. Such a recalibration and model revision should lead to a model that has greater predictive power and reliability.

Discussion

Whether the errors in this case were significant in relation to the overall problem can be answered best (or perhaps, only) by those who sponsored the model study in light of (1) what they expected, (2) what actions were taken or not taken because of these predictions, and (3) what predictive alternatives were available. The model predictions represented only one hypothesis of future contaminant spreading. The 1980 test drilling was designed, to a large extent, to test that very prediction. The process of collecting data is most efficient when guided by an objective of hypothesis testing. A major value of the model so far has been to help optimize the data collection and monitoring process; that is, the predictive model offers a means to help decide how frequently and where water samples should be collected to track the plume. Thus, modeling and data collection are an iterative process.

Tucson Airport

Background

Ground water within a zone approximately 6 mi long and 1 mi wide in the vicinity of the Tucson Airport is contaminated with organic solvents, primarily trichloroethene (Figure 5.9) (40 CFR §261.24). The contaminated ground water is in an extensive alluvial aquifer that Tucson uses as its principal aquifer. The Tucson area, with a population of 517,000, is one of the largest metropolitan areas in the country that is totally dependent on ground water for drinking water, and the trichloroethene contamination was viewed as a threat to the integrity of the water supply system. The area containing the contaminated ground water is listed on the National Priority List and is known as the Tucson Airport Area Superfund Sites (51 Fed. Reg. 21,054 [1986]).

Several potential sources of the ground water contamination were identified in a remedial investigation conducted at the site (Rampe, 1985). The available data indicated that several industrial facilities had used trichloroethene in their processes, and that industrial wastewaters had been disposed of in ponds and drainage ditches and

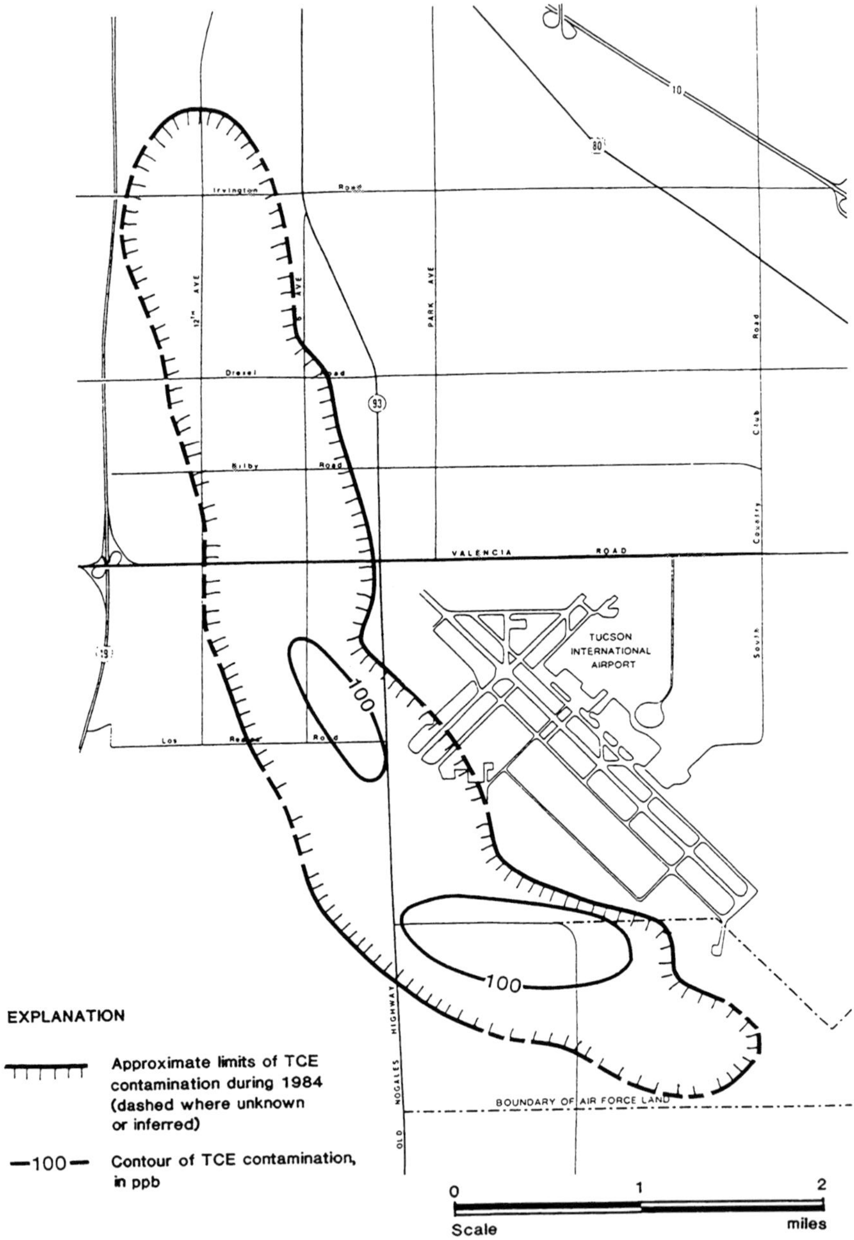

FIGURE 5.9 Location of Tucson Airport. SOURCE: Adapted from CH2M Hill, 1987.

on the ground. There were no data, however, to indicate how much trichloroethene had been lost to the subsurface at any individual facility.

The remedial investigation made some general conclusions regarding the significance of individual source areas, but these conclusions were questioned by the potentially responsible parties (Arizona Department of Health Services, 1986). As a result, EPA, Region IX, asked CH2M Hill, an environmental consulting firm, to conduct an assessment of potential sources. Specifically, CH2M Hill was requested to do the following:

- Assess the possibility of contribution to the ground water contamination from the various potential sources that had been identified.
- Assess the ranges of relative contributions for each potential source and the probability distribution associated with the range.

This discussion of contamination in the Tucson Airport area is largely extracted and paraphrased from the draft report by CH2M Hill (1987) and the remedial investigation prepared for the Arizona Department of Health Services by Schmidt (1985) and Mock et al. (1985).

Hydrogeologic Setting

The Tucson Airport is located within the upper Santa Cruz basin, an alluvial basin bordered by north to northwest trending fault block mountains (Fenneman, 1931). Basin fill deposits, predominantly sands, sandy gravels, and clayey sands, make up the aquifer system in the vicinity of the Tucson Airport. Three distinct aquifer units are identified in the area: an upper coarse-grained unit that extends to a depth of about 200 ft, a middle fine-grained unit that is about 100 ft thick, and a lower unit with lenses of coarse-grained materials whose total thickness is unknown. Ground water contamination is generally confined to the upper unit, which consists mainly of clayey sands interbedded with lenticular deposits of sand and sandy gravels.

Water levels in the upper unit are currently about 100 ft below land surface, and the water-level gradient is toward the north-northwest. Water levels in the upper aquifer fell about 30 ft from 1952 to 1981, but they have been relatively stable since 1981, possibly as a result of reductions in pumpage by the city of Tucson in this area (Mock et al., 1985). Ground water levels in the lower unit differ from those in the upper unit by 60 to 100 ft (CH2M Hill, 1987).

Approach

The approach used to address the objectives of this study consists of two aspects: (1) establishing a basis for making assessments of the relative contributions to ground water contamination from multiple sources and (2) developing estimates of the contributions that each source had made to areas of ground water contamination. Models were used to make each of these assessments, and these models are described below.

Model to Assess Relative Contributions

The contribution of a source to contamination of an aquifer can be assessed either by evaluating the quantity of contaminants contributed to the aquifer from a source or by evaluating the area of the aquifer affected by the release of contaminants. This study started with the assumption that, for purposes of assigning responsibility for cleanup of an aquifer, it is appropriate to assess relative contributions from sources based on areas of contamination. The reasons given for this assumption were the following:

- the extraction system required to contain or withdraw contaminated ground water is not dependent on the levels of contamination but, rather, is nearly directly proportional to the area of contamination;
- treatment costs are influenced more by volume of water treated than by actual levels of contamination; and
- the quantity of contaminants released cannot be reliably estimated when only low levels of contamination are observed.

The relative contribution of a source area was assessed with the following equation:

$$RC_a = \left(\frac{1}{A_T} \sum_{i=1}^{m} \frac{A_i}{L_i} \right) \times 100,$$

where

RC_a = percent relative contribution of source a to the area of contaminated aquifer

A_i = area of aquifer where source *a* had contributed contaminants, a subarea of the contaminated aquifer
L_i = number of sources that contributed contaminants to area A_i
m = number of subareas contaminated by source *a*
A_T = total area contaminated

The use of this model to assess relative contributions is illustrated in Figure 5.10. In the example given, three sources, *a*, *b*, and *c*, contributed contamination. Source *a* contributed to the entire area, source *b* contributed to one-fourth of the area, and source *c* contributed to one-half of the area. The relative contributions to the contamination of the entire area for sources *a*, *b*, and *c* are 71, 8, and 21 percent, respectively. The relative contributions calculated for each source take into consideration those sources that affect an area that has also been affected by other sources. This procedure provides a method for considering overlapping contributions from multiple sources.

Model of Source Contributions

The area of contamination resulting from a contaminant release from an individual source could not be determined with the information available on the distribution of contaminants. Rather, the area of contamination from a source was estimated using a two-dimensional numerical contaminant transport model. The flow transport model consisted of two separate steps: first, a finite-difference ground water flow model was developed and calibrated; then, a transport model based on the method of characterizations (Konikow and Bredehoeft, 1978) was used to estimate contaminant spread in the aquifer. In developing the flow model, CH2M Hill first divided the aquifer, on the basis of aquifer test data, into seven zones of equal permeability. Then an automatic parameter estimation technique, based on the method described by Cooley (1977), was used to refine the permeability estimates in each zone. The refined estimates were those that minimized the sum of the squared difference between the simulated levels and the observed 1984 water levels.

Once a steady-state flow model was calibrated to simulate 1984 conditions, stream lines were calculated from each source area. Because there was good agreement between the distribution of contamination and the pattern of stream lines, it was concluded that a steady-state flow model is a reasonable representation of the ground

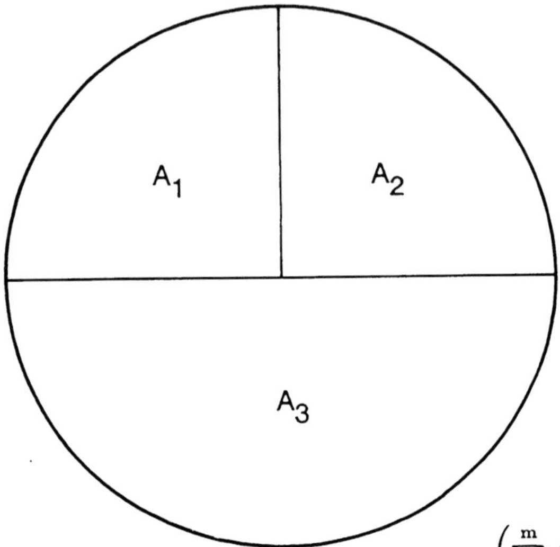

INPUT:

Source a contributed to subareas A_1, A_2, A_3

Source b contributed to subarea A_1 only

Source c contributed to subareas A_1 and A_2

A_T = 1 = area of affected aquifer under consideration

$$RC_a = \left(\sum_{i=1}^{m} \frac{A_i}{L_i}\right) \frac{100}{A_T}$$

$$= \left(\frac{A_1}{3} + \frac{A_2}{2} + \frac{A_3}{1}\right) \frac{100}{A_T}$$

$$= \left(\frac{0.25}{3} + \frac{0.25}{2} + \frac{0.5}{1}\right) \frac{100}{A_T}$$

$$= 0.71 \; \frac{100}{A_T}$$

$$= 71\%$$

$$RC_b = 8\%$$

$$RC_c = 21\%$$

FIGURE 5.10 Example calculation of relative contribution to aquifer contamination.

water flow field and therefore that source releases could be simulated by using the steady-state flow field and transient transport.

Prior to simulating mass transport, it was necessary to estimate the quantity of trichloroethene released from each source and the timing of the releases. The timing of the releases was estimated on the basis of historical records of trichloroethene usage, and the quantity released was estimated on the basis of the quantity of trichloroethene in the aquifer. The effects of potential source releases were simulated

TABLE 5.3 Calculated Estimates of Trichloroethene Releases from Source Areas

Source Area	Total Trichloroethene Release (gal.)	Time of Release
A	400	1952–1977
B	155	1952–1984
C	155	1952–1984
D	55	1964–1984
E	130	1952–1984

for each source acting alone, and in addition, the combined effects of the potential sources were evaluated for various release and timing scenarios. A trial and error procedure was used to estimate the most probable release scenario.

Results and Conclusions

The results indicated the potential for a large area of contamination to develop from rather small amounts of trichloroethene, and the release scenario described in Table 5.3 was judged to be most representative of conditions observed in the field.

The individual plumes, calculated by simulating each source independently, are shown in Figure 5.11a, and the combined plume is shown in Figure 5.11b. On the basis of these simulated plumes, the relative contribution of each source to the total contamination north and south of Los Reales Road was calculated using the relative contribution equation discussed above. The calculated relative contributions are shown in Table 5.4.

CH2M Hill noted that the model simulations provided results for only specific cases: that is, the best estimate of permeability in each zone and one contaminant release scenario. CH2M Hill stated that if multiple simulations were performed to take into account the plausible variations in permeability and trichloroethene releases, a range of relative contributions could be developed for each source in a quantitative manner. They noted, however, that the estimates of uncertainty about the release would be subjective owing to lack of data. Therefore only a qualitative assessment was made of the range of relative contributions.

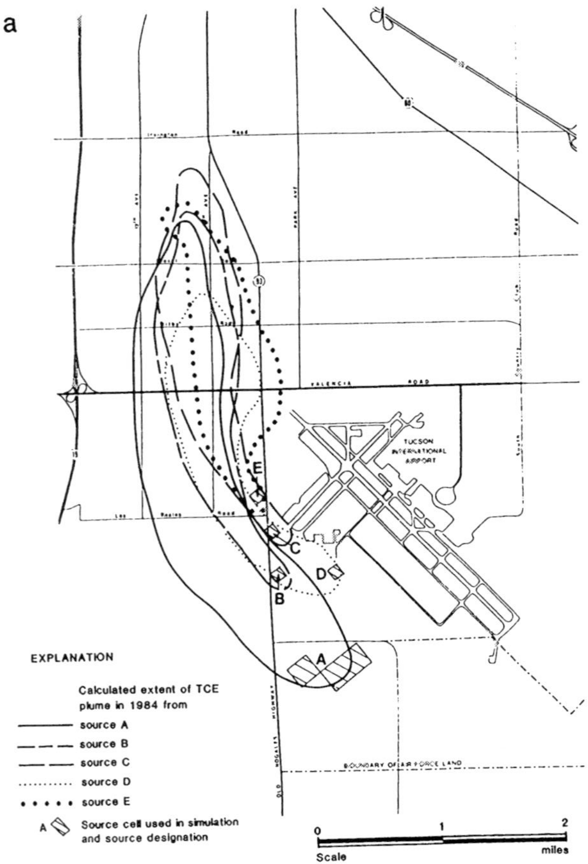

FIGURE 5.11a Simulated individual trichloroethene plumes. SOURCE: Adapted from Hill, 1987.

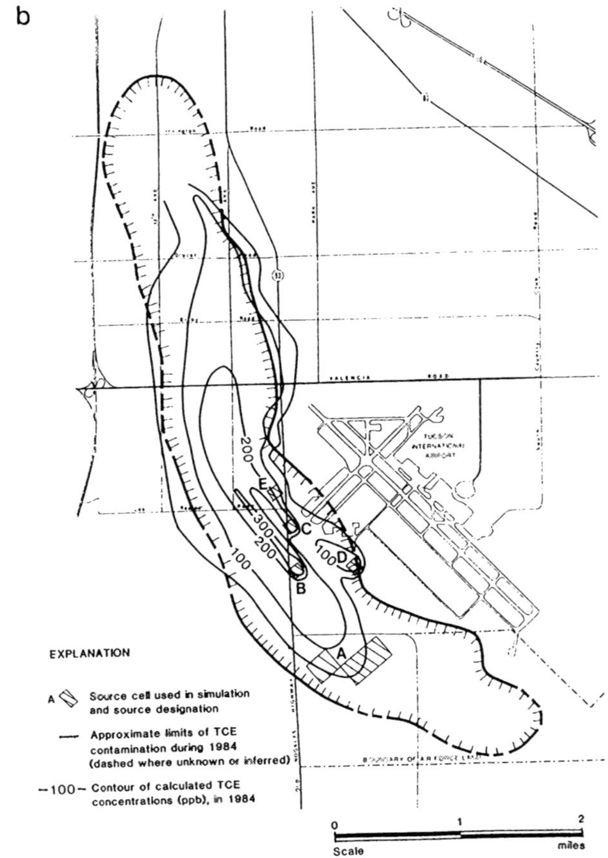

FIGURE 5.11b Simulated combined trichloroethene plumes. SOURCE: Adapted from Hill, 1987.

TABLE 5.4 Calculated Relative Contributions from Individual Source Areas

Source	Relative Contribution (percent)
South of Los Reales Road	
A	74
B	4
C	3
D	19
E	0
North of Los Reales Road	
A	33
B	14
C	20
D	11
E	22

Regulatory Context

Ground water contamination in the Tucson Airport area was discovered in the early 1950s. Intensive investigations of ground water contamination did not begin until 1979, when a sampling program was initiated at the request of EPA. In March 1981 an investigation conducted by EPA under the authority of the Comprehensive Environmental Response, Compensation, and Liability Act (CERCLA) identified trichloroethene and chromium contamination in the ground water. As a result, seven municipal wells were removed from service, and the site was listed on the National Priority List.

In November 1982, Hughes Aircraft Company and the U.S. Air Force assumed responsibility for investigating contamination south of Los Reales Road, while EPA assumed responsibility for investigating contamination north of Los Reales Road (Figures 5.11a and 5.11b). Hughes and the Air Force concluded that the contamination beneath the Air Force property was caused by past disposal of waste solvents and claimed that no continuing sources existed on the facility. Consequently, the Installation and Restoration Program (IRP) conducted by the Air Force in 1985 did not contain any proposed source control measures. The Tucson Airport Area Remedial Investigation, which was managed by the Arizona Department of Health under a cooperative agreement with EPA, was concluded in 1985. To date, over 100 monitoring wells have been drilled to identify, characterize, model, and monitor the contamination in the area (Environmental Protection Agency, 1988a).

Eighteen remedial alternatives were designed and analyzed for the Air Force property. Ground water extraction and recharge were chosen as the preferred remedy. This remedy for the Air Force property, which began operation in 1988, involves pumping, treating, and recharging approximately 26 billion gal. of water over a 10-yr period (Environmental Protection Agency, 1988a). This remedy is described in the Record of Decision issued for the site on July 25, 1988. No remedy has yet been selected for the area north of Los Reales Road. The modeling study conducted by CH2M Hill for EPA that attempted to assign liabilities was severely criticized by several of the potentially responsible parties, and as a result no agreement has yet been reached on an appropriate remedial action for this area.

Discussion

Ground water models are frequently used to determine sources of observed contamination. In general, the information on the current distribution of contaminants and hydrogeologic conditions is insufficient to allow a unique solution for the location of sources and the timing of source releases. This is particularly true when all potential sources are located along the same stream line and there are no marker chemicals for a specific source. Ground water models, however, can be used to help set bounds on the range of possible contributions from individual sources.

S-Area, Niagara Falls, New York

Background

The S-Area landfill is located on the southeast corner of Occidental Chemical Corporation's Buffalo Avenue Plant in Niagara Falls, New York. Approximately 63,100 tons of chemical waste was deposited at the site. The S-Area landfill is one of four landfills in the Niagara Falls, New York, area that were operated by Occidental Chemical Corporation (OCC), formerly known as Hooker Chemicals Plastics Corporation. The other landfills are Love Canal, Hyde Park, and the 102nd Street landfill.

Ground water flow and transport models have been used extensively at all of these sites, and the use of these models is particularly well documented (Mercer et al., 1985; C. Faust, affidavits in Civil Action Nos. 79-988 and 79-989 in the U.S. District Court for the Western District of New York, 1984 and 1985, respectively).

These contaminant transport models have also been incorporated into legally enforceable documents and have been evaluated and approved by a court. For simplicity, this case study focuses primarily on the models used at the S-Area site, because they illustrate the complex processes that can be simulated with the current generation of ground water transport models.

A major concern at the landfill is discontinuities in an underlying confining bed that allow dense nonaqueous-phase liquids (NAPLs) to contaminate a bedrock aquifer. The chemicals in the landfill will be contained after remediation by an integrated system of barrier walls, plugs, drains, and a cap that is designed to prevent off-site migration. A conceptual hydrogeologic cross section of the landfill before and after remediation is shown in Figure 5.12. Prior to remediation, hydraulic gradients are downward, and ground water and NAPL flow into the bedrock where the clay and fill are missing. After remediation, the drains, walls, and cap on the site are intended to create a sufficient upward hydraulic gradient to reverse the flow of ground water and NAPL into the bedrock.

A one-dimensional, two-phase flow model was developed by Arthur D. Little, Inc. (ADL) to establish what upward hydraulic gradient would prevent downward migration of NAPL at the S-Area landfill (Arthur D. Little, Inc., 1983; Guswa, 1985; C. Faust, Affidavit in Civil Action No. 79-988 in the U.S. District Court for the Western District of New York, 1984 (particularly paragraphs 42-44)). The model considers, among other things, the effects of lithology-dependent capillary pressure functions, hydraulic gradients, and permeability variations. Subsequently, a two-dimensional, two-phase flow model was developed by EPA's consultant to ensure that the one-dimensional model was appropriate for selecting a remedy for the site. After the initial remedies were selected for the site, a three-dimensional model was developed and is currently being used to evaluate conditions at the site and the potential effectiveness of additional remedies at the site. The model's use to design the remedy is discussed in this case study.

Site Conditions

The NAPL found at S-Area has a specific gravity of approximately 1.5 and consists primarily of trichlorobenzene, tetrachlorobenzene, pentachlorobenzene, tetrachloroethylene, hexachlorocyclopentadiene, and octachlorocyclopentene (S. Fogel, Affidavit in Civil Action No. 79-988 in the U.S. District Court for the Western District

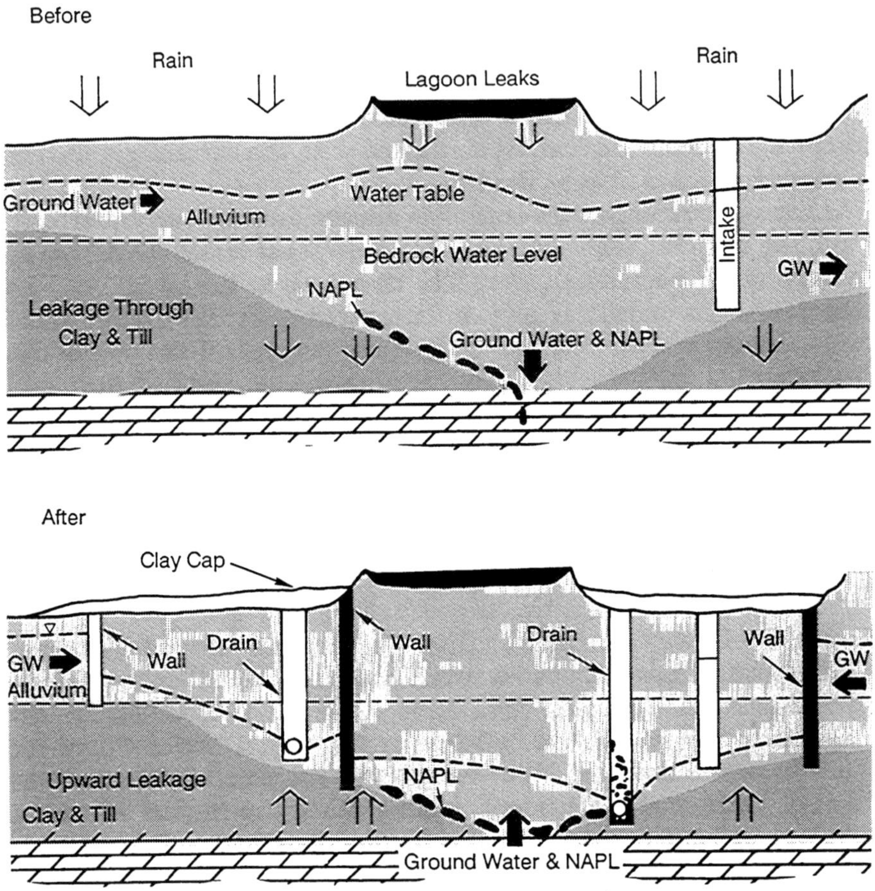

FIGURE 5.12 A conceptual cross section of the hydraulic containment system to be implemented at the S-Area landfill. SOURCE: Cohen et al., 1978.

of New York, 1984). These liquids have been observed in discrete discontinuous zones in the landfill. Geologic logs indicate a lithologic contact between unconsolidated glacial deposits and bedrock (Lockport dolomite) at an elevation of about 541 ft. The base of the unconsolidated glacial deposits is a clay ranging in thickness from about 0.25 to 15 ft. The clay is overlain by a relatively thick (up to 16 ft) fine sand layer containing scattered zones of silt and fine gravel. This is overlain by about 14 ft of artificial fill. Bedrock water-level measurements indicate a potentiometric elevation of about 561 ft. Water levels measured in the overlying unconsolidated deposits indicate a positive head difference between the overburden and the underlying bedrock of between 2 and 5.5 ft. Under these conditions therefore, a vertical downward flow component exists.

Model Formulation

The two models developed to design the remedies used the method of finite differences. The ADL model employed the implicit pressure-explicit saturation (IMPES) method to solve the two coupled equations of flow for an immiscible nonaqueous phase and water. The air phase is neglected. The ADL model also used a mesh-centered grid, whereas the other model, referred to as SWANFLOW (simultaneous water and NAPL flow), used a block-centered approach (GeoTrans, 1985).

To evaluate the potential for downward NAPL flow, a vertical column 23 ft long was divided into 24 blocks (nodes). The model was constructed with a 2-ft negative head difference (downward flow) between the water table and bedrock potentiometric level. The domain contains three different porous materials. The upper 20 ft consists of a fine sand with a hydraulic conductivity of 10^{-5} cm/s ($k = 1.02 \times 10^{-14}$ m^2). The fine sand is underlain by 1 ft of clay ($K = 10^{-7}$ cm/s; $k = 1.02 \times 10^{-16}$ m^2). The clay is underlain by the Lockport dolomite bedrock ($K = 10^{-3}$ cm/s; $k = 1.02 \times 10^{-12}$ m^2). The residual saturation values for water and NAPL were assumed to be 20 and 10 percent, respectively. Other simulation data are given in Tables 5.5 and 5.6.

Results and Conclusions

The results show that a barrier to downward migration of NAPL is provided by capillary pressure differences between the sand and clay (Figure 5.13). This condition has been confirmed in recent field investigations at the S-Area site (Faust and Guswa, 1989).

A comparison between the results of the two numerical models is shown in Figure 5.13. The saturations calculated by SWANFLOW and the ADL code at approximately 250 days are shown. The results from the two models compare favorably; however, there are some differences, especially just above the clay layer. The differences are probably caused by some combination of instability in the IMPES technique, alternative gridding and time steps used in the two codes, and slight differences in the relative permeability relationships (the ADL [1983] model provided for hysteresis in capillary pressure).

TABLE 5.5 Capillary Pressure and Relative Permeability Data for ADL Simulation 1

Capillary Pressure (N/m^2)[a]	Water Saturation	Relative Permeabilities	
		Water	NAPL
Fine sand and bedrock			
103,425.0	0.00	0.00000	1.00000
103,425.0	0.10	0.00000	0.82000
103,425.0	0.20	0.00000	0.68000
27,580.0	0.30	0.04000	0.55000
10,343.0	0.40	0.10000	0.43000
7,585.0	0.50	0.18000	0.31000
7,447.0	0.60	0.30000	0.20000
7,309.0	0.70	0.44000	0.12000
7,171.0	0.80	0.60000	0.05000
7,033.0	0.90	0.80000	0.00000
6,895.0	1.00	1.00000	0.00000
Clay			
206,850.0	0.00	0.00000	1.00000
206,850.0	0.10	0.00000	0.82000
206,850.0	0.20	0.00000	0.68000
165,480.0	0.30	0.04000	0.55000
134,453.0	0.40	0.10000	0.43000
110,320.0	0.50	0.18000	0.31000
93,082.0	0.60	0.30000	0.20000
82,740.0	0.70	0.44000	0.12000
75,845.0	0.80	0.60000	0.05000
72,398.0	0.90	0.80000	0.00000
68,950.0	1.00	1.00000	0.00000

[a] N = newton (i.e., kg-m/s^2).

SOURCE: Arthur D. Little, Inc., 1983.

TABLE 5.6 Data Used in ADL Simulation 1

Parameter	Value
Porosity	0.2
Permeability	
Fine sand	1.02×10^{-14} m^2
Clay	1.02×10^{-16} m^2
Bedrock	1.02×10^{-12} m^2
Density of water	1,000 kg/m^3
Density of NAPL	1,500 kg/m^3
Water viscosity	0.001 kg/m-s
NAPL viscosity	0.001 kg/m-s
D_z (vertical dispersion length)	0.3048 m

SOURCE: Arthur D. Little, Inc., 1983.

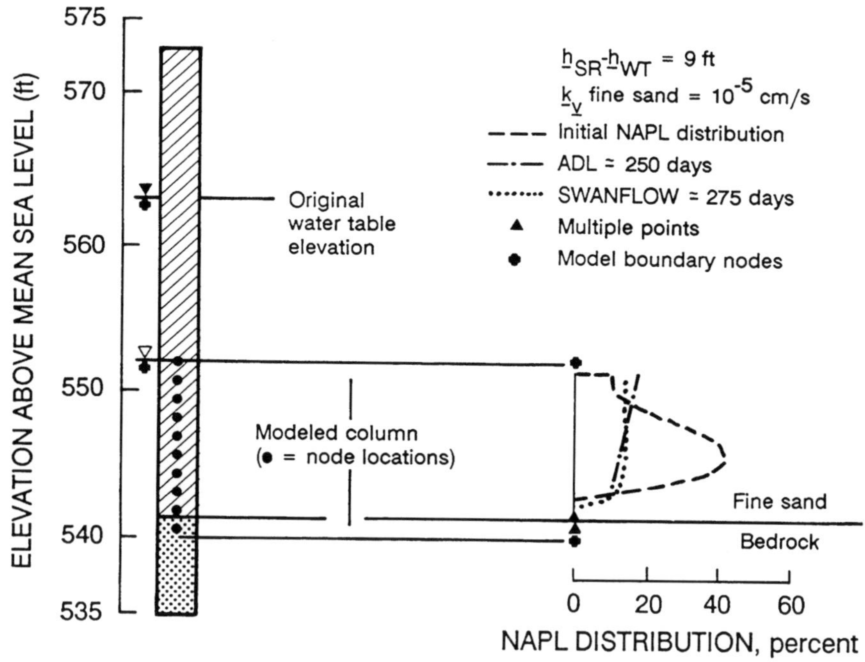

FIGURE 5.13 NAPL saturation profiles at one time for the two-layer simulation.

The effects of a water-phase hydraulic gradient on NAPL migration were also examined via these simulations, where the clay layer was assumed to be missing. As shown in Figure 5.13, the results of this series of simulations indicated that a minimum upward head difference of 9 ft between the water table elevation and bedrock potentiometric level in the vicinity of a clay layer discontinuity could be sufficient to prevent downward migration of NAPL into the bedrock (Guswa, 1985). This figure indicates NAPL saturations at about 250 days. As shown, there is a noticeable upward movement of NAPL.

Data have been collected as part of a remedy designed to lower the hydraulic head in the overburden sand. These data will be used to confirm the remedy as well as modeling results.

Regulatory Context

In December 1979 the federal government filed four lawsuits to obtain cleanup of four OCC landfills. EPA, the state of New

York, and OCC negotiated an extensive set of remedies. The S-Area Consent Decree incorporated these remedies, including the one-dimensional, two-phase containment transport model discussed above. The consent decree was lodged with the court on January 10, 1984.[23] Consent decrees are subject to a 30-day public comment period. If the consent decree is adequate, proper, and in the public interest, the Department of Justice and the court finally approve it (see 28 CFR §50.7). In this particular case, the province of Ontario requested and the court granted a formal evidentiary hearing to review the consent decree. These models were subject to close, critical scrutiny during the public comment period and court hearing, including scrutiny by consultants hired by the province of Ontario.[24] The court held that the consent decree was "fair, adequate, and consistent with public policy . . . [and] will adequately protect the public interest in health and the environment."[25]

Two consultants employed by EPA, as well as EPA and state personnel, oversaw and peer-reviewed the development of the ADL model (G. Pinder, Affidavit in *United States* v. *Hooker Chemicals and Plastics ("S" Area Landfill)*, Civil Action No. 79-988 in U.S. District Court for the Western District of New York, 1984, particularly paragraphs 23-25). The two-dimensional, two-phase flow model was developed by one of EPA's consultants to ensure that the one-dimensional model was appropriate at the site (C. Faust, Affidavit in Civil Action No. 79-989 in the U.S. District Court for the Western District of New York, 1985).

Discussion

This case study illustrates the use of relatively complex models of ground water and NAPL flow to help design a remedial action for a hazardous waste site. Field studies have shown that both of these models were able to simulate observed field conditions. The results of these model studies have demonstrated that the current generation of ground water models can be used to investigate the migration of an immiscible, denser than water fluid within an aquifer. Interestingly, this study also shows that a one-dimensional model can be just as useful as a two-dimensional model in the investigation of the appropriateness of a remedial action.

NOTES

1. 42 USC §9601 et seq. and 40 CFR Part 300. National Oil and Hazardous Substances Pollution Contingency Plan, 53 Fed. Reg. 51,394 (1988), contains the proposed new Superfund regulations.

2. *United States* v. *Mottolo,* 605 F. Supp. 898, 902 (DNH 1985).

3. Chemical Carcinogens; A Review of the Science and Its Associated Principles, 1985, 50 Fed. Reg. 10,372 (1985).

4. Guidelines for Estimating Exposures, 51 Fed. Reg. 34,042 (1986).

5. Ibid.

6. Hazardous Waste Management System; Identification and Listing of Hazardous Waste; Final Exclusion and Final Organic Leachate Model (OLM).

7. Hazardous Waste Management System; Identification and Listing of Hazardous Waste.

8. Hazardous Waste Management System; Identification and Listing of Hazardous Waste; Notification Requirements; Reportable Quantity Adjustments.

9. Hazardous Waste Management System; Land Disposal Restrictions.

10. Hazardous Waste Management System; Identification and Listing of Hazardous Waste; New Data and Use of These Data Regarding the Establishment of Regulatory Levels for the Toxicity Characteristic; and Use of the Model for the Delisting Program.

11. Underground Injection Control Program: Hazardous Waste Disposal Injection Restrictions; Amendments to Technical Requirements for Class I Hazardous Waste Injection Wells; and Additional Monitoring Requirements Applicable to All Class I Wells.

12. Citing D. Morganwalp and R. Smith, 1987, Modeling of Representative Injection Sites, EPA report in progress.

13. This discussion of contamination in the Tucson Airport area is largely extracted and paraphrased from the report by CH2M Hill (1987) and the Remedial Investigation prepared for the Arizona Department of Health Services by Schmidt (1985) and Mock et al. (1985).

14. Hazardous Waste Management System; Identification and Listing of Hazardous Waste: Use of a Generic Dilution/Attenuation Factor for Establishing Regulatory Levels and Chronic Toxicity Reference Level Revisions.

15. Hazardous Waste Management System; Identification and Listing of Hazardous Waste.

16. Hazardous Waste Management System; Identification and Listing of Hazardous Waste; Notification Requirements; Reportable Quantity Adjustments.

17. *McLouth Steel Products Corp.* v. *Thomas,* 838 F.2d 1317, 1320 (D.C. Cir. 1988).

18. Hazardous Waste Management System; Identification and Listing of Hazardous Waste.

19. Hazardous Waste Management System; Identification and Listing of Hazardous Waste; Final Exclusion Rule.

20. Hazardous Waste Management System; Identification and Listing of Hazardous Waste; Final Denials.

21. See supra, note 10.

22. *The Washington Post,* p. 1 (October 9, 1981).

23. *United States* v. *Hooker Chemicals & Plastics Corp. (S-Area),* Civ. Act. No. 79-988 (filed January 10, 1984).

24. *United States* v. *Hooker Chemicals & Plastics Corp.*, 607 F. Supp. at 1061.
25. Ibid. at 1070.

BIBLIOGRAPHY

Arizona Department of Health Services. 1986. Responsiveness Summary, Results of the Tucson Airport Area Remedial Investigation, 15 pp.

Arthur D. Little, Inc. 1983. S-Area to Phase Flow Model. Prepared for Wald, Harkrader & Ross (now merged with Pepper, Hamilton & Scheetz), Washington, D.C.

CH2M Hill. 1987. Assessment of the Relative Contribution to Groundwater Contamination from Potential Sources in the Tucson Airport Area, Tucson, Arizona. Prepared for U.S. EPA Region IX.

Cohen, R. M., R. R. Rabold, C. R. Faust, J. O. Rumbaugh III, and J. R. Bridge. 1978. Investigation and hydraulic containment of chemical migration: Four landfills in Niagara Falls. Civil Engineering Practice (Spring), 33–58.

Cooley, R. L. 1977. A method for estimating parameters and assessing reliability for models of steady-state ground-water flow, 1. Theory and numerical properties. Water Resources Research 13, 318–324.

Domenico, P. A., and V. V. Palciauskas. 1982. Alternative boundaries in solid waste management. Ground Water 20, 301–311.

Downey, J. S., and E. J. Weiss. 1980. Preliminary Data Set for Three-Dimensional Digital Model of the Red River and Madison Aquifers. U.S. Geological Survey Open-File Report 80-756, Denver, Colo.

Environmental Protection Agency. 1986a. Superfund Public Health Evaluation Manual. OSWER Directive No. 9285.4-1, Washington, D.C.

Environmental Protection Agency. 1986b. RCRA Ground-water Monitoring Technical Enforcement Guidance Document. OSWER Directive No. 9950.1, Washington, D.C.

Environmental Protection Agency. 1987a. Alternate Concentration Limit Guidance Part 1: ACL Policy and Information Requirements. Interim Final. OSWER Directive No. 9481.00-6C, EPA/530-SW-87-017. Washington, D.C.

Environmental Protection Agency. 1987b. Evaluation of Risk-Based Decision-making in RCRA. Annotated Briefing. Internal document, p. 5.

Environmental Protection Agency. 1988a. Evaluation of Hughes Aircraft, U.S. Air Force Plant No. 44, Tucson, Ariz. EPA/700-8-87-037, Hazardous Waste Ground-Water Task Force, Washington, D.C.

Environmental Protection Agency. 1988b. Final Review Draft Guidance on Remedial Actions for Contaminated Ground Water at Superfund Sites. OSWER Directive No. 9283, Washington, D.C., pp. 3–22.

Environmental Protection Agency. 1988c. Selection Criteria for Mathematical Models Used in Exposure Assessments: Ground-water Models. EPA/600/8-88/075. Washington, D.C.

Faust, C. R., and J. H. Guswa. 1989. Simulation of three-dimensional flow of immiscible fluids within and below the unsaturated zone. Submitted to Water Resources Research. In press.

Fenneman, N. M. 1931. Physiography of the Western United States. McGraw-Hill, New York.

GeoTrans, Inc. 1985. SWANFLOW: Simultaneous Water, Air, and Nonaqueous Phase Flow, Version 1.0. Documentation prepared for Environmental Protection Agency.

Grove, D. B. 1977. The Use of Galerkin Finite-Element Methods to Solve Mass-Transport Equations. U.S. Geological Survey Water Resources Investigation 77-49, 55 pp.

Guswa, J. H. 1985. Application of Multi-Phase Flow Theory at a Chemical Waste Landfill, Niagara Falls, New York. Pp. 108-111 in Proceedings of the Second International Conference on Groundwater Quality Research, published by the National Center for Ground Water Research, Stillwater, Okla.

Konikow, L. F. 1976. Preliminary Digital Model of Ground-Water Flow in the Madison Group, Powder River Basin and Adjacent Areas, Wyoming, Montana, South Dakota, North Dakota, and Nebraska. U.S. Geological Survey Water Resources Investigation 63-75, 44 pp.

Konikow, L. F., and J. D. Bredehoeft. 1978. Computer Model of Two-Dimensional Solute Transport and Dispersion in Ground Water: Techniques of Water-Resources Investigations of the United States Geological Survey. Book 7, Chapter C2, U.S. Geological Survey, 90 pp.

Lewis, B. D., and F. J. Goldstein. 1982. Evaluation of a Predictive Ground Water Solute-Transport Model at the Idaho National Engineering Laboratory, Idaho. U.S. Geological Survey Water Resources Investigation 82-55, 71 pp.

Mercer, J., C. R. Faust, R. M. Cohen, P. F. Andersen, and P. S. Huyakorn. 1985. Remedial action assessment for hazardous waste sites via numerical simulation. Waste Management and Research 3, 377–387.

Mock, P. A., B. C. Travers, and C. K. Williams. 1985. Results of the Tucson Airport Area Remedial Investigation, Volume III, Contaminant Transport Modeling. Arizona Department of Water Resources, 106 pp.

Rampe, J. 1985. Results of the Tucson Airport Area Remedial Investigation, Phase I, Volume III, Evaluation of the Potential Sources of Groundwater Contamination near the Tucson International Airport. Arizona Department of Health Services.

Reeves, M., D. S. Ward, N. D. Johns, and R. M. Cranwell. 1986a. Data Input Guide for SWIFT II; the Sandia Waste-Isolation Flow and Transport Model for Fractured Media. Release 4.84. NUREG/CR-3162, U.S. Nuclear Regulatory Commission, Washington, D.C.

Reeves, M., D. S. Ward, N. D. Johns, and R. M. Cranwell. 1986b. Theory and Implementation for SWIFT II; the Sandia Waste-Isolation Flow and Transport Model for Fractured Media. Release 4.84. NUREG/CR-3328, U.S. Nuclear Regulatory Commission, Washington, D.C.

Reeves, M., D. S. Ward, P. A. Davis, and E. J. Bonena. 1987. SWIFT II Self-Teaching Curriculum; Illustrative Problems for the Sandia Waste-Isolation Flow and Transport Model for Fractured Media (revised). NUREG/CR-3925, U.S. Nuclear Regulatory Commission, Washington, D.C.

Robertson, J. B. 1974. Digital Modeling of Radioactive and Chemical Waste Transport in the Snake River Plain Aquifer at the National Reactor Testing Station, Idaho. U.S. Geological Survey Open-File Report ID0-22054, 41 pp.

Ross, B., J. W. Mercer, S. D. Thomas, and B. H. Lester. 1982. Benchmark Problems for Repository Siting Models. NUREG/CR-3097, U.S. Nuclear Regulatory Commission, Washington, D.C.

Schmidt, K. D. 1985. Results of the Tucson Airport Area Remedial Investigation, Vol. I, Summary Report. Arizona Department of Health Services, 113 pp.

Science Advisory Board. 1984. Report on the Review of Proposed Environmental Standards for the Management and Disposal of Spent Nuclear Fuel, High Level and Transuranic Radioactive Wastes (40 CFR §191). Report to the EPA by the High-Level Radioactive Waste Subcommittee.

Siefken, D., G. Pangburn, R. Pennifill, and R. J. Starmer. 1982. U.S. Nuclear Regulatory Commission Low Level Waste Licensing Branch Technical Position—Site Suitability, Selection, and Characterization. NUREG-0902, U.S. Nuclear Regulatory Commission, Washington, D.C.

Silling, S. A. 1983. Final Technical Position or Documentation of Computer Codes for High-Level Waste Management. NUREG-0856, U.S. Nuclear Regulatory Commission, Washington, D.C.

Trescott, P. C., G. F. Pinder, and S. P. Larson. 1976. Finite-Difference Model for Aquifer Simulation in Two Dimensions with Results of Numerical Experiments. Techniques of Water-Resources Investigations of the U.S. Geological Survey, Book 7, Ch. Cl. U.S. Government Printing Office, Washington, D.C., 116 pp.

U.S. Nuclear Regulatory Commission. 1982. Final Technical Position on Documentation of Computer Codes for High Level Waste Management. NUREG-0856, Washington, D.C., p. 2.

U.S. Nuclear Regulatory Commission. 1988. Standard Review Plan for the Review of a Licensed Application for a Low Level Radioactive Waste Disposal Facility: Safety Analysis Report 1988, NUREG-1200, Revision 1. Washington, D.C.

Wilkinson, G. F., and G. E. Runkle. 1986. Quality Assurance (QA) Plan for the Computer Software Supporting the U.S. Nuclear Regulatory Commission's High-Level Waste Management Program. NUREG/CR-4369, U.S. Nuclear Regulatory Commission, Washington, D.C.

Woodward-Clyde Consultants, Inc. 1981. Well-Field Hydrology Technical Report for ETSI Coal Slurry Pipeline Project. U.S. Bureau of Land Management.

Zamuda, C. 1986. The Superfund record of decision process: Part 1—The role of risk assessment. Chemical Waste Litigation Reporter 11, 847.

6
Issues in the Development and Use of Models

INTRODUCTION

In the United States, two agencies, the U.S. Nuclear Regulatory Commission (USNRC) and the Environmental Protection Agency (EPA), are particularly concerned with ground water modeling to support many of their regulatory activities. Their experience with the uses of models has been completely different. The USNRC, while placing considerable emphasis on developing guidance for the selection and use of models, has never really employed them for regulatory purposes.

The USNRC's low-level waste (LLW) program has yet to be tested, because no applications for licenses have been received by the USNRC. In any case, the USNRC is likely to receive fewer than 10 applications for disposal sites. The high-level radioactive waste program is also untested. License applications for high-level waste repositories have not been received, and none are expected before 1995.

The EPA's experience in using models is documented to a much greater extent because of the number of active sites under its jurisdiction. Models play an important role in EPA-related activities; however, many problems related to the use of models have emerged. For example, prior reviews of the Superfund cleanup process have concluded the following:

- all analytical methodologies suffer from a lack of knowledge on the fundamental process underlying observed phenomena (National Research Council, 1988);
- models do not account for all the processes affecting the fate and impact of the contaminants (National Research Council, 1988);
- models lack accuracy when confronted with a high degree of heterogeneity (complex hydrogeology, multiple contaminants, two-phase flow, and variable susceptibility in populations) (National Research Council, 1988);
- there is no clear guidance provided by agencies concerning when to use and how to select models (International Ground Water Modeling Center, 1986; Office of Technology Assessment, 1982);
- the decision concerning when to use a model and which code to use is often left to the discretion of the contractor who was hired by EPA or a potentially responsible party (International Ground Water Modeling Center, 1986);
- there is limited understanding among EPA staff concerning which models are available (International Ground Water Modeling Center, 1986);
- there is inadequate expertise within federal and state regulatory agencies to apply such models (Office of Technology Assessment, 1982);
- the validity of some codes for the problem to which they are applied has not been established (Office of Technology Assessment, 1982);
- EPA enforcement offices strongly discourage the use of proprietary models (International Ground Water Modeling Center, 1986; Office of Technology Assessment, 1982);
- there is inadequate quality assurance, quality control, and peer review (Office of Technology Assessment, 1982); and
- there is a reluctance to use models if their use would be considered controversial (Office of Technology Assessment, 1982).

The committee's review confirmed most of these findings. The problem is not a lack of appropriate documents to guide the modeling process. One can see from the list that the basic problems concern the lack of training and experience in the people who are choosing and using models, deficiencies or limitations in the codes themselves, and scientific barriers that determine to what extent models are able to incorporate relevant processes. The committee addresses these issues in this chapter.

THE PEOPLE PROBLEM

It should be apparent from earlier chapters outlining the state of the science that modeling ground water flow and contaminant transport is not a trivial exercise. Ideally, a modeler should have a broad background in earth sciences with particular strengths in hydrogeology, low-temperature geochemistry, and analytical and numerical mathematics. This background will have developed through graduate and undergraduate studies and will have been tempered by relevant experience. A significant problem in dealing with regulatory agencies is the lack of individuals who are trained at an appropriate level to understand and use models.

For example, EPA's ground water and contaminant transport modeling needs currently outpace its actual use of models in virtually all program areas (Office of Technology Assessment, 1982). EPA currently has an insufficient number of qualified and experienced hydrogeologists and other professionals knowledgeable in contaminant transport modeling (Office of Technology Assessment, 1982). Superfund hydrogeologists are quitting their jobs at a rate 6 times higher than the average for other federal government employees (General Accounting Office, 1987). The more experienced hydrogeologists are leaving EPA at a higher rate than the younger professionals, and the situation is likely to become worse. Most states possess even more limited capabilities (Council of State Governments, 1985; Environmental Protection Agency, 1987; General Accounting Office, 1987; International Ground Water Modeling Center, 1986). The substantial increase in the need for site-specific regulatory decisions in all the EPA programs concerned with regulating ground water can only exacerbate the breadth and depth of these shortages and critical needs.

Contaminant transport models simply cannot be used unless people who are experts in ground water processes and models are available to select, apply, and peer-review such models. As a result, EPA's system for selecting and applying models is guaranteed to result in misuses of such models.

The solution to this problem will require (1) recruiting and retaining more qualified and experienced personnel; (2) establishing specific guidelines and criteria for the use of contaminant flow models; (3) instituting peer review techniques; and (4) providing technical assistance and additional training.

The lack of qualified individuals in the regulatory agencies at all levels has had some significant ramifications. For example, some of the methods chosen to expedite hazardous waste cleanups are contrary to good science. EPA's policy of performing remedial investigations in less than 6 months often provides insufficient data for a complete characterization of the site. Given the seasonal character of ground water flow, it would be prudent to measure water levels over a longer time frame.

The rush to judgment on Superfund remedies risks more than a "bad" scientific decision or an economically wasteful cleanup. Decisions based on inadequate data may aggravate a problem or, at least, prolong its eventual remedy. The prudent, if not necessary, course of action in such cases is to proceed in orderly phases, such as in the S-Area case study. The committee recognizes the desirability of, and public mandate for, expediting hazardous waste cleanups. Those components of the remedial action for a site that reasonably can be taken with limited data—for example, interception and treatment of the ground water plume—should be implemented immediately. Other components of a site remediation plan can be implemented at a more measured pace once the primary potential source of exposure is eliminated.

Another problem generated through inexperience is an overreliance on the results of a modeling exercise. Computer models have a unique capacity to appear more certain, more precise, and more authoritative than they really are. As a result, assumptions, even wholly unrealistic ones, can be stated with deceptive precision and seeming accuracy by being included in a computer model.

Special care therefore must be taken in presenting the results of such modeling. Decisionmakers (whether they be heads of agencies, judges, or juries) must understand the distinction between scientific fact and science policy. If policy is relied on to make a decision, the policy rationale should be explicitly identified.

Faced with the problem of an overall lack of qualified staff to use models and interpret results, regulatory agencies have a natural tendency toward simplification through the use of standard models and worst-case assumptions, as is done in the hazardous waste delisting program (Environmental Protection Agency, 1987; International Ground Water Modeling Center, 1986). This decision is motivated by a concern about the lack of adequate resources and a preference for using overprotective assumptions. There is an inherent conflict between using more complex, site-specific models and using simpler models; i.e., "[s]tandardization may increase consistency, but tends to trade off accuracy, producing answers that are not always appropriate in all situations" (Environmental Protection Agency, 1987).

A site-specific decision should be based on the actual conditions existing at a site. More certainty should be required if the output of the model is used directly to trigger additional regulatory action than if the model is used as an interpretative tool to better understand how contaminants migrate near the site.

The committee recognizes the need to follow the mandate of the enabling statutes, use health-protective assumptions, and consider the practical limitations on agency resources. The committee, however, believes that the use of standard models at specific sites lacks a scientific basis. The use of overly simplistic models, such as the vertical-horizontal spread (VHS) model, at Superfund sites or other hazardous waste sites (1) would be an arbitrary distortion of the remedial selection process, (2) could reduce protection of the public health by misallocating finite cleanup resources, and (3) would result in the imposition of substantial costs with no commensurate environmental or public health benefit.

The Environmental Protection Agency's choice of remedies can also be affected by the choice of model and the assumptions used in such a model. For example, EPA may use an advection-dispersion contaminant transport model to predict the future concentrations of chemicals at local drinking water wells to derive the on-site soil cleanup levels (i.e., soil levels that would not result in off-site ground water concentrations above health-based ground water cleanup levels) (Record of Decision, July 1985, McKin Site, Maine, RO1-85-009) or to estimate the time that it will take to achieve various cleanup levels by alternative remedial actions (Record of Decision, August 1985, Old Mill, Ohio, RO5-85-018; Record of Decision, September 1987a, Suffern Well Field, New York, RO2-87-042). Such a model will not take into account dilution, adsorption, volatilization, or biodegradation and other more realistic features (Record of Decision, September 1987b, Rose Township, Michigan, RO5-87-052). For example, an advection-dispersion model generally will overestimate the concentration and underestimate the travel time for the contaminants, thus making the problem appear much more serious than it is in reality.

Extreme worst-case assumptions can drive the remedy selection process toward draconian and extremely costly remedies. The selection of these assumptions as input to models is also prone to misuse (*Pesticide and Toxic Chemical News*, 1987). The difference between worst-case assumptions and levels predicted by contaminant transport modeling (no less real) can be substantial.

The benefit of using extreme worst-case assumptions is often simply administrative convenience to the agency; i.e., using such assumptions eliminates the need to obtain additional data and make difficult expert judgments. This benefit must be weighed against the additional cost or the possibility that the assumption will significantly underestimate the risk. Worst-case assumptions should never be preferred over actual data. Some assumptions may be so unrealistic that their use is inappropriate.

UNCERTAINTY AND RELIABILITY

Modeling can be defined as the art and science of collecting a set of discrete observations (our incomplete knowledge of the real world) and producing predictions of the behavior of a system. Such predictions will be necessarily uncertain, as will be our knowledge of the true behavior of the system. The goal of this section is to identify and discuss the scientific, technical, and practical issues that arise in applying models to particular sites, and to develop procedures and guidelines to help assure that these issues are addressed during the model application process. A convenient framework for organizing a discussion of uncertainty and reliability in modeling is presented by Figure 6.1. What is shown is one possible representation of the process of applying a model to a regulatory (or other) decisionmaking problem. This representation rests on the assumption that the ultimate goal of a modeling exercise is a prediction of the behavior of the real world. That is, there is a "true" system, made up of the geologic environment (the soils and/or aquifers), climatic stresses (precipitation and evaporation), subsurface flora and fauna, and human-induced stresses (e.g., irrigation wells). The success of a modeling exercise will depend on the degree to which the model prediction agrees with the behavior of this true system. Therefore the reference in discussing and/or assessing the accuracy of the modeling process is this real system, indicated by the top path of Figure 6.1.

The state and characteristics of the real world may be described by a set of information termed the inputs to the system, such as the spatial distribution of soil and aquifer properties, or the time histories of system stresses. These inputs are often highly variable in time and/or space. Some may be inherently uncertain, such as future time series of rainfall infiltration and subsequent recharge.

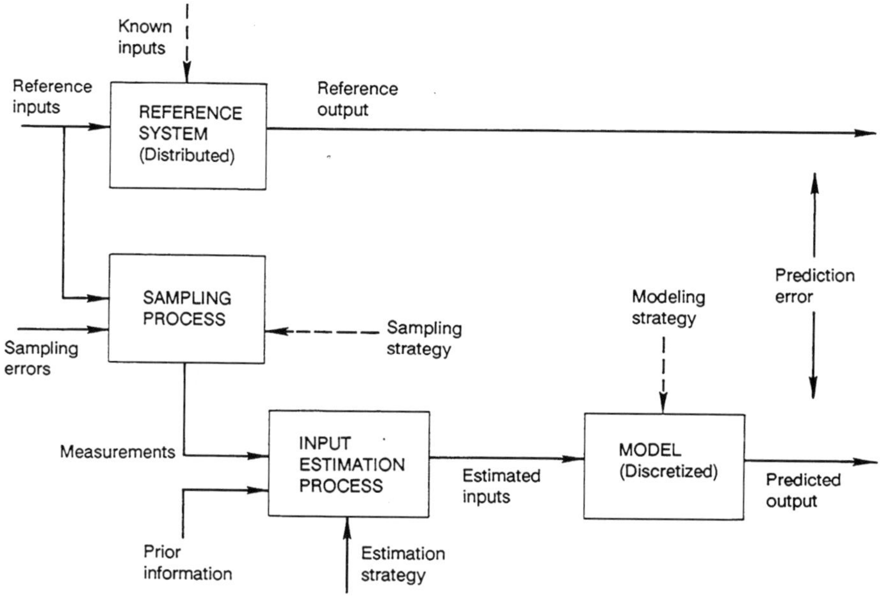

FIGURE 6.1 Conceptual framework for ground water model accuracy analysis. SOURCE: McLaughlin and Wood, 1988a.

The processes at work in the real system, including those induced by proposed management actions, act on these inputs to yield the true, or real, outputs that characterize the behavior of the system. Such outputs might be contaminant concentration distributions in space and time, travel times, mass losses, or exposure levels and durations at selected locations. These true outputs are, of course, themselves often variable and uncertain. Even though ground water flow and transport systems tend to smooth out the variability of inputs, much variability and uncertainty remain in the true outputs. The following sections use this conceptual model to describe the major sources of uncertainty in the modeling process.

The Sampling Process

One can observe the real world only via a sampling process. We make a finite number of observations, choosing what parameters to measure, how to measure them (what instrument to use), where to measure them, and when to measure them. In other words, a sampling scheme is designed and implemented. For example, one might collect a set of cores during well drilling and measure the permeability of subsamples of each core in the laboratory using a permeameter. Alternatively, one might collect a set of water samples from wells and analyze each in the laboratory for contaminant concentration. Such a sampling scheme typically provides a set of discrete quantitative observations of one or more parameters of interest, or sometimes more continuous, qualitative information about the system (e.g., the geologic sedimentary environment).

A sampling process introduces uncertainty. First, the measurement process itself introduces uncertainty in the form of instrument errors. Every measurement device has associated with it a measurement error. Such errors usually contain a random (uncertain) component (and are often biased). Second, the sampling process introduces uncertainty because of incomplete information. The system can be observed only at a small set of points, and conditions between sampling points are not known with certainty, whether in space or time or both. This uncertainty is obviously most critical for systems characterized by significant spatial and temporal variability. Thus the real system is an uncertain one because of (1) its inherent randomness, (2) measurement error, and perhaps most important, (3) limited sampling of the highly variable physical, chemical, and biological properties of ground water systems. This uncertainty applies to both the inputs and the outputs of the system. All modeling is conducted without certain knowledge of the true state of a ground water environment. The magnitude of our uncertainty is a function of the spatial heterogeneity and temporal variability of aquifer properties, boundary conditions, dependent variables, the density of observation points relative to the scale of the variability, and the measurement techniques. With these general concepts in mind, we can address more specific issues concerned with field sampling and data collection.

Field sampling, experimental design, and related data analysis issues are topics that have not traditionally received much attention from ground water modelers. While most modelers appreciate the need for good field data, they have often had to depend on others for the data used in their models. Published field data have typically been taken at face value and have been freely extrapolated and generalized beyond their original purpose. This situation has begun to change, partly as a result of the demanding requirements of hazardous waste studies and partly because modelers are beginning to take a broader view of the modeling process, which recognizes that data issues need to be taken seriously.

As mentioned earlier in this report, ground water systems are difficult to observe and describe, not only because they are hidden from view, but also because they are three-dimensional and often very heterogeneous. Hydrogeological properties observed at one location may give relatively little information about conditions only a few meters away. Soil strata or rock fractures only a few centimeters thick may greatly influence the movement of water and contaminants but pass undetected in a typical field survey. Such heterogeneities limit our ability to generalize from laboratory measurements to field conditions or from one site to another. The ground water sampling problem is complicated further by the expense of well drilling, which is still the primary method used to gain information about subsurface flow and transport. Drilling is time-consuming and labor intensive, and requires specialized equipment. Moreover, the drilling process disturbs the subsurface environment and, as a result, compromises the accuracy of pump tests and contaminant data collected from observation wells. Although alternative sampling methods based on geophysical or remote sensing technology have been applied successfully in some situations, they are generally even less reliable than well samples. The expense, difficulty, and inaccuracy of field sampling all tend to have an adverse impact on ground water modeling. Most modeling studies must make do with a very limited amount of unreliable data, which at best give only a rough picture of actual subsurface conditions. This basic fact needs to be recognized in any realistic assessment of the prediction capabilities of ground water models.

Generally speaking, the field data used to estimate the inputs and check the predictions of ground water flow models are compiled from historical hydrogeologic surveys that were not planned with modeling in mind. Examples include periodic status reports issued by irrigation districts and state water agencies (primarily in the western United States), U.S. Geological Survey (USGS) water supply papers and open file reports, and water resource atlases compiled by a number of different governmental agencies. Until recently, many of the data included in these surveys were collected by local well drillers and geologists concerned primarily with water supply. These data tend to cover regions that are larger than those of interest in ground water contamination studies and therefore rarely deal with local geologic or hydrologic anomalies that may control transport in the vicinity of a hazardous waste site. In most hazardous waste studies, these traditional data sources are useful only for defining the boundary conditions of a site-specific flow model.

The field data used in contaminant transport models typically have a very different history from those used in flow models. Most contaminant concentration measurements are collected at or near a contaminated site after an indication that some problem exists (e.g., observations of unusual taste or odor in well water). These measurements are usually limited and scattered, reflecting the locations of existing water supply wells rather than the geometry of the contaminant plume (or plumes). Furthermore, contaminant data may be even more difficult to interpret than hydrogeologic data because the compounds observed and their physical state depend on chemical and biological conditions in the subsurface environment (see Chapter 2).

These comments suggest that there will be a need for a specialized problem-oriented sampling program at most hazardous waste sites. Because sampling resources are nearly always quite limited, the objectives of the sampling program need to be spelled out carefully so that a systematic and cost-effective field strategy can be developed. This strategy needs to be flexible enough to be able to deal with unanticipated results and unforeseen logistic problems but specific enough to provide guidance to drilling crews and managers responsible for approving budget expenditures. The dichotomy of flexibility and specificity is one that arises time and again in practical sample programs.

A site-specific hazardous waste field sampling program may have many different objectives, which can exert conflicting demands on limited resources. Some frequently encountered objectives include the following:

- assessment of the severity of a newly discovered contamination problem (i.e., a reconnaissance study);
- monitoring of a known but more or less controlled hazardous waste site (e.g., for enforcement of a consent decree);
- monitoring of the performance of a remediation strategy (e.g., a pumping, treatment, and reinjection system); and
- acquisition of data needed to develop or test a predictive model.

Because this report is primarily concerned with ground water modeling, the focus is on the last of these objectives. It should be noted, however, that modelers may need to reconcile their needs with those of other data users competing for limited resources and, in the process, may be forced to make compromises and adjustments in their approach.

Recently, there has been a significant increase in research on the design of model-oriented ground water monitoring programs (Chu et al., 1987; Graham and McLaughlin, 1989a,b; Knopman and Voss, 1987, 1988; McLaughlin and Wood, 1988a,b). Although the specific methods proposed differ considerably, they generally view data collection as a way to reduce uncertainty. If it is possible to relate a specific data collection strategy to the uncertainty inherent in modeling, then it is possible to compare different strategies and select the one that is, in some sense, the best.

Field sampling studies can, at least in principle, help to reduce the major types of uncertainty including (1) lack of knowledge about the processes that control contaminant transport and transformation at a particular site and (2) incomplete knowledge of the spatially and temporally variable environmental factors that influence these processes. In fact, it is useful to divide a model-oriented sampling program into two phases: the first (less structured) phase attempts to identify relevant transport processes, whereas the second (more specific) phase attempts to quantify heterogeneous hydrogeologic and biochemical properties. Each of these is briefly discussed below.

Process and Parameter Identification

There is no truly systematic way to identify the physical, chemical, and biological processes at work at a particular contaminated site. This is a difficult scientific and engineering problem that requires creativity and experience as well as a good ability to identify inconsistencies and suspicious anomalies in a limited set of observations. Nevertheless, it is possible to state three general principles that may help structure the field studies needed to support a site-specific model development effort.

1. A site-specific description of contaminant transport is strongly dependent on the quality of the flow model used to develop estimates of subsurface water velocities. Considerable care should be used in developing the inputs to the flow model, particularly in reference to the following:

- Well logs and water level data should be examined to determine the importance of three-dimensional (vertical) effects related to geological stratification, density differences, buried sources, and so on. Vertical homogeneity should not be assumed without supporting documentation.
- Borings and surficial geological information should be used to identify the primary hydrogeological features of the site including, as much as possible, local anomalies that may influence contaminant migration.
- Boundary conditions should be used to match the local flow field with known regional patterns and to account for interactions with surface features such as lakes or streams. Some sampling resources should be reserved for gathering information about flow boundary conditions, including recharge from the surface, if this information is not already available.
- The average value and likely range of soil properties such as hydraulic conductivity, porosity, and specific storage should be estimated from pump and piezometer tests, grain-size analyses, and if possible, permeameter tests of soil samples. A range of tests should be used so that variations at different scales can be assessed.
- If flow in the unsaturated zone or through highly conductive fractures is important, special care should be taken to assess, at least in a qualitative way, the role of these features. It is risky to assume that such effects are unimportant just because they are inconvenient.

2. The sampling program should recognize that contaminant dispersion is largely a manifestation of unknown hydrogeological heterogeneities (see Chapter 2). These heterogeneities produce a more tortuous (heterogeneous) subsurface flow field than would be obtained under uniform conditions. Although intentional or after-the-fact tracer analyses can be used to estimate macroscopic dispersivities, it is also possible to derive these macrodispersivities from theoretical analyses that recognize the variable nature of the small-scale flow field (Dagan, 1984; Gelhar and Axness, 1983; Neuman et al., 1987). This is an important and somewhat controversial issue, which at least deserves consideration when designing a field sampling program.

3. The sampling program should attempt to either verify or rule out the various chemical and biological processes that may play a role in the transport and transformation of contaminants at the site. Because the responsible mechanisms depend largely on the chemical composition of the contaminants, the field program should provide a waste inventory, if one is not already available. The role of processes such as sorption, precipitation and colloidal transport, biodegradation, and multiphase transport and volatilization should be assessed before any detailed modeling is undertaken. This is a difficult task that is not readily codified but can benefit greatly from experience and from familiarity with the scientific literature on the transformation of subsurface contaminants. Particular care should be taken to ensure that sampling procedures and analytical techniques do not, by their very nature, automatically rule out observation of a potentially important transformation process.

These general guidelines suggest that a significant portion of a model-oriented sampling effort should be devoted to a somewhat unstructured exploratory study that identifies the dominant processes to be included in subsequent modeling efforts. Many of the data collected in this exploratory phase can later be used to estimate the value of key model inputs.

Input estimation and validation, the second phase of a model-oriented field sampling program, presume that the processes included in the model are, in fact, the ones that control contaminant behavior at the site of interest. The sampling program should provide the data needed to obtain the most accurate model inputs and predictions possible, subject to resource constraints. This is, in fact, a statement of the traditional sampling problem addressed by classical statistics (see, for example, Cochran and Cox, 1957; Cox, 1958; Federov, 1972; Kiefer and Wolfowitz, 1959). Much of the literature dealing with this problem is based on simple regression models and is oriented toward controlled field experiments (e.g., agricultural evaluations of hybrid crop types). Useful variants on the traditional approach are found in the extensive literature on the design of rain and stream gage networks (International Association of Hydrological Sciences, 1986) and in the geostatistical literature, which is largely concerned with mapping heterogeneous soil properties (Delhomme, 1979; Journel and Hiujbregts, 1978). Sampling design for contaminant transport applications is a new and active research area that has yet to produce practical techniques for designing site-specific monitoring programs.

Nevertheless, it is possible to state some general principles that are beginning to emerge:

- If a sampling program is intended to provide data for estimating model inputs, it should be designed to minimize an appropriate measure of estimation uncertainty. In practice, this measure is often the mean-squared estimation error, although many other measures have been proposed. Because the estimation error depends on the structure of the model (e.g., the computational grid used to define model inputs) and on the input estimation procedure, sampling design should be viewed as an integral part of the modeling process. In particular, the structure (e.g., resolution) of the model should reflect field sampling constraints, and the sampling program should be designed to serve the model. This is a simple but frequently ignored principle that needs to be given more attention in practical modeling applications.
- It is probably neither realistic nor desirable to seek a unique "optimal" sampling design, i.e., one that is unequivocally better than all competitors. This is because formalized optimization cannot consider all the factors that influence the selection (and evolution) of a given design. Such factors include logistic and legal constraints (e.g., access), conflicting objectives, unanticipated interruptions and delays, uncertainty about the relative importance of various natural processes, and the ever-present possibility of a totally unexpected discovery partway through the sampling process. Instead of seeking an optimum, an attempt should be made, at any stage of sampling, to identify the best among a set of reasonable alternatives.
- The unpredictable nature of field sampling (which, after all, is most informative when it yields the least predictable results) suggests that practical sampling programs should evolve sequentially, with resources committed over a series of stages rather than all at once. Thus the results of each stage of sampling are used to update the models that form the basis for the sampling design. In fact, if field data suggest that a particular model is inappropriate, it may be discarded altogether and replaced by a more appropriate one before additional resources are committed.
- Sampling designs are highly dependent on the technical capabilities and cost of the sampling devices used to collect data in the field and on the methods used to obtain and preserve samples for later analysis in the laboratory. Sampling technology is changing rapidly; therefore, a range of alternatives should be carefully considered before extensive resources are committed to specialized equipment. It is

probably best to use a mix of several different approaches, some well-established and some more experimental (depending on the scope and objectives of the field effort). Remote sensing techniques may, for example, provide useful qualitative information about regional water level trends but not give a good match to observations obtained from more localized (and more expensive) piezometer measurements. A judicious combination of both techniques might be the best approach at some sites.

These guidelines, like the ones stated earlier, confirm that the design of model-oriented field sampling programs is still a largely ad hoc endeavor that requires a good understanding of subsurface physical, chemical, and biological processes, of model and input estimation algorithms, and of sampling technology. It is rare for any one individual to be capable in all of these areas, which makes sampling design a truly multidisciplinary effort that typically requires the active participation of several specialists. This situation is likely to continue for the foreseeable future.

Input Estimation

In order to apply a chosen model formulation to a particular site, certain model inputs are required. These include coefficients, such as hydraulic conductivity, specific storage, porosity, and thermodynamic constants; boundary conditions, such as aquifer geometry and piezometric head, contaminant concentration, and mass flux along or across the aquifer boundary; and initial conditions, such as head and concentration distribution at a particular point in time. The most appropriate values of these inputs depend not only on the true physical, chemical, and biological state of the real world, but also on the amount of aggregation or averaging in the model formulation (e.g., the size of grid elements) and the model structure. The process of selecting appropriate input values is termed "input estimation."

Many of the coefficients and input variables included in ground water models must be estimated on a case-by-case basis, usually from a relatively limited number of field observations of related quantities. Input estimation is one of the most difficult, and often most frustrating, aspects of ground water modeling. Engineers and decisionmakers who use models need to understand the difficulties inherent in the estimation process if they are to make informed judgments about the desirability of modeling and the accuracy of model predictions.

It is useful to distinguish at the outset two types of model inputs, which are treated somewhat differently in practice:

1. Constitutive Coefficients and Parameters. When a ground water model is formulated from basic principles (such as conservation of mass or conservation of energy), quasi-empirical "laws" are often used to relate certain model variables. Important examples include Darcy's law, which relates specific discharge to the hydraulic local head gradient, and Fick's law, which relates dispersive flux to the local concentration gradient (see Chapter 2). Such empirical laws introduce various so-called constitutive parameters that are generally not directly observable, but, rather, must be inferred from observations of other model variables. These include parameters such as hydraulic conductivity, dispersion coefficients, and partition coefficients. Field studies indicate that many of these parameters vary dramatically over space and, possibly, over time. Saturated hydraulic conductivity variations can easily vary 3 or 4 orders of magnitude over the scale of a typical contaminant site (Dagan, 1986; Gelhar, 1986). Unsaturated conductivities vary even more, reflecting their dependence on moisture content. Theoretical and experimental analyses indicate that field-scale dispersivity coefficients can vary over time and with the scale of the experiment, sometimes by orders of magnitude. The nonobservability and variability of constitutive parameters make them difficult to estimate, particularly when field measurements are limited.

2. Forcing Terms and Auxiliary Conditions. Most ground water models include forcing terms, which account for sources and sinks of water or dissolved contaminants. Flow models typically include pumping and recharge terms, whereas transport models typically include terms that describe where and when contaminants are introduced into the subsurface environment. In some cases, such forcing terms are measured directly. In other cases, they are inferred from measurements of more accessible variables, or they are simply postulated (as, for example, when the effects of a proposed cleanup strategy are being investigated). Forcing terms generally act in the interior of a simulated region, at wells or disposal sites. Ground water and associated contaminants can also enter or leave the region across boundaries. The boundary conditions imposed on a model's solution can have an important impact on predicted flow and transport behavior. Parameters included in these boundary conditions (such as specified heads, concentrations, and fluxes) can sometimes

be inferred from field observations. They are more often simply postulated. Similar remarks apply to initial conditions, which can be important in transient simulations.

The traditional approach to ground water input estimation, developed largely in water resource investigations of large aquifers, focuses on constitutive parameters such as hydraulic conductivity and dispersivity. In aquifer-scale applications, it is often feasible to select model boundaries and simulation periods so that auxiliary conditions can be readily specified. This is why, for example, boundaries are often drawn along flow divides (yielding no-flux boundary conditions) and simulations are often initialized when the ground water system is at steady state (enabling the initial conditions to be computed rather than measured). The traditional approach may not always work in hazardous waste applications, where the scale of the modeling problem is often much smaller (e.g., hundreds of meters rather than tens of kilometers) and where background contaminant concentrations may be highly uncertain. In such cases, it is more realistic to view boundary and initial conditions as inputs that need to be estimated from field measurements.

Methods for estimating ground water inputs vary greatly, depending on the application and the resources available to the modeler. Input estimation is often posed as a so-called inverse problem. That is, model inputs are estimated from measurements of the model's outputs (the "inverse" of the direct modeling problem that computes outputs from specified inputs). The concept of "model calibration" is a variant on inverse estimation. Calibration is the process of adjusting model inputs until the resulting predictions give a reasonably good fit to observed data. This process, which sounds reasonable enough on the surface, has the disadvantage of being "ill posed" in most ground water applications. The ill-posedness arises from the fact that an infinite number of input combinations can generally provide acceptable fits to historical measurements. These combinations of parameters may differ greatly and may give significantly different results when used to predict future conditions. Ill-posedness (or nonuniqueness) has been studied by a number of researchers in various areas of science and engineering and is a problem that is familiar to most modelers.

The primary practical solution to ill-posedness problems in ground water model calibration is to use "prior information" to guide or constrain parameter adjustments. Such information includes data obtained from soil samples, well and piezometer tests, laboratory experiments, and tracer tests, as well as sound engineering and geological judgment based on experience with similar sites. If prior information is available, a variety of automated procedures may be used to carry out the inverse estimation process. Good reviews are provided by Carrera and Neuman (1986) and Yeh (1986). Generally speaking, such procedures are used mostly by researchers, although they are beginning to be applied more frequently by the USGS and by some consulting firms. In the future, it is likely that automated inverse estimation algorithms will be included as part of the modeling packages distributed for general use by practicing hydrogeologists.

The measurements used to estimate both constitutive parameters and auxiliary conditions are typically obtained at discrete times and locations (usually at monitoring wells). These local measurements need to be extrapolated over larger regions if they are to be used for modeling purposes. There are many important examples. Well observations of hydraulic head (water level) and solute concentrations constitute the primary source of data for model calibration. These observations need to be contoured to provide a synoptic picture of the desired model response. Head and concentration contour maps are also needed to define source terms and auxiliary conditions that may be difficult to estimate with inverse techniques. Maps of hydraulic conductivities and porosities deduced from soil samples, piezometer tests, or geophysical measurements provide a good source of prior information for use in model calibration efforts.

A number of methods are available for estimating regional distributions of ground water model inputs from scattered well observations. One of the most popular is a least-squares procedure known as kriging (Delhomme, 1979; Journel and Hiujbregts, 1978). This procedure, which originated in the mining industry, provides estimates of the accuracy of the contours it generates. Estimation accuracy depends, as might be expected, on the distance from observation points and on the heterogeneity of the contoured variable. Some care must be taken in using the procedure, however. It can give particularly deceptive pictures of contaminant plumes if used in its standard form, which assumes that contaminant concentration is in isotropic (nondirectional) and stationary (homogeneous) random fields. In reality, contaminant concentrations are highly anisotropic and nonstationary, particularly near sources. Generally speaking, traditional kriging packages are useful for contouring soil properties and other smoothly varying quantities but should not be used to extrapolate contaminant concentrations beyond sample locations.

Recently, several researchers have attempted to combine aspects of traditional inverse estimation with aspects of kriging, to provide a more integrated approach to input estimation (see, for example, Hoeksema and Kitanidis, 1984). While this approach is a worthwhile endeavor, most applied hydrogeologists will continue, at least for the near future, to estimate model inputs in a more or less ad hoc way, using trial-and-error adjustments based on contoured data and intuition, with occasional help from an automated package. Whether the estimation procedure is manual or automated, the final results will depend greatly on how well the modeler understands the factors that relate predictions to the model's structure and input values. In particular, four issues should be kept in mind:

1. Model Formulation and Structure. The success of any parameter estimation effort is critically dependent on the validity of the underlying model formulation. If the model's structure ignores important sources, geological heterogeneities, physical processes, or chemical reactions, parameter estimation will be reduced to a fitting exercise that forces available inputs to compensate (usually inadequately) for an improper formulation.

2. Past Versus Future Performance. A good fit to historical data does not guarantee good predictions, particularly if the historical fit is based on a small amount of data or if it does not test model capabilities that are required for making predictions. It is dangerous to "overfit" historical measurements by adjusting parameters beyond reasonable ranges. Although historical fits can reveal important information about model behavior, they should be related to other relevant factors, including qualitative geological observations.

3. Sensitivity Analysis. Sensitivity analysis provides a useful (although not perfect) way to identify the model inputs that have the most influence on model predictions, at least over a specified range. Although a detailed sensitivity analysis can be laborious and time-consuming, it is usually feasible to carry out a small-scale exploratory analysis that focuses on a few critical inputs identified, most likely, by informed intuition. The sensitivity analysis should guide the selection of inputs included in the estimation process (see item 4).

4. Choice of Estimated Inputs. The results of a ground water input estimation depend greatly on which inputs are based on field data and which are assumed to be well known. If, for example, a velocity used in a transport model points in the wrong direction, it

will not be possible to obtain correct predictions by adjusting dispersion coefficients or retardation rates, no matter how sophisticated the estimation algorithm is. When in doubt, all important inputs, e.g., source locations, source magnitudes, and auxiliary conditions, should be included in the estimation process.

Underlying all of these points is the theme of model accuracy introduced earlier in this chapter. Input estimation is one of several interrelated factors that influence the accuracy of a model's predictions. If the subsurface environment is very heterogeneous, measurements are very limited, or the model is improperly formulated, it is unlikely that the estimation process will be able, by itself, to ensure accurate predictions. The effort devoted to input estimation, and the sophistication of the estimation procedure, should be judged in a larger context that includes data collection and model formulation.

Model Validation and Accuracy Assessment

The output of a model application exercise is a set of data representing the predicted behavior of the ground water system in response to one or more proposed management actions. These predictions are determined by the particular combination of sampling process, model formulation, input estimation, and solution technique employed. They depend on decisions made in each step of the model application process.

The accuracy (or validity) of a particular model application should logically be measured by the magnitude of the prediction errors, i.e., by some measure of the difference between the response of the real world and the response of the simulated system to management actions. Such a comparison is complicated by the fact that prediction errors are uncertain because of sampling error. Also, the scale and/or level of aggregation of both the real and the modeled system response must be consistent if a valid comparison is to be made. It is particularly difficult to develop a priori assessments of modeling accuracy. Traditional methods of accuracy assessment focus on comparisons of predictions to historical measurements, evaluating goodness-of-fit after the fact. Although important indicators of model performance, such methods do not truly measure prediction errors.

Model validation is a term that means different things to different people, largely because it is rarely defined with any precision. This general concept has both technical and policy origins. From a technical viewpoint, modelers feel a need to confirm or verify the hypotheses used in their models. This is generally accomplished by comparing predictions to observations, preferably under controlled conditions that can clearly reveal the sources of any discrepancies. From a policy viewpoint, regulatory agencies, courts, and public officials feel a need for standards that can be used to certify the results of a modeling effort or, more narrowly, to certify the use of a particular computer program. In this case, the implicit goal seems to be to reduce the risk that a model will lead to inappropriate decisions. Although this risk clearly depends on the model's accuracy, it also depends on how the model is used, i.e., on how the decision is made.

Checking a model's validity by comparing its predictions with measurements is an important part of classical statistics. There are many statistical tests for evaluating models and related hypotheses. Traditional statistical methods are not particularly useful in ground water modeling studies, however, for several reasons. First, there are rarely enough measurements in ground water applications to provide a statistically rigorous test of a model's explanatory capabilities. These measurements are typically available at scattered well locations, which are spaced further apart than characteristic scales of variability. Second, the conditions prevailing when the measurements were collected may not reflect those that the model is designed to simulate. Finally, most classical statistical tests are based on assumptions that are not necessarily met in complex subsurface environments. These tests typically assume that the model's structure is perfect, and they are based solely on an analysis of the effects of measurement error. In reality, natural heterogeneity and deficiencies in model structure are likely to be far more important than measurement error.

Since rigorous statistical validation tests are generally not appropriate in ground water applications, model validation is typically an ad hoc exercise that does not have a firm scientific foundation. Instead, model parameters are adjusted until a "reasonable" fit is obtained and the result is presented as a "validated model." Modelers practically never declare their models to be "invalidated," primarily because ground water models nearly always have enough adjustable parameters to fit a limited set of field observations. This leads us to ask how we can distinguish a good fit that is based on artificial manipulation of an overparameterized model from a good fit that is based on an accurate description of the processes that control contaminant transport.

One way to respond to the question posed above is to extend or generalize the concept of model validation. Instead of focusing on whether a model is valid, one can focus on evaluating its accuracy. That is, one can attempt to quantify the probability that the model's predictions deviate from reality by more (or less) than a specified amount at any given time or location. Accuracy can be conveniently expressed in terms of confidence limits or, given appropriate assumptions, in terms of the risk associated with a particular decision based on model predictions. Such information is ultimately both more useful and more realistic than a certification that a model is or is not validated.

Although a quantitative assessment of model accuracy would undoubtedly be useful, it is not clear how such an assessment can be developed when the data needed to test model performance are very limited. One approach to this dilemma is to carry out a model "error analysis." Prediction errors can ultimately be traced to three basic sources:

1. natural heterogeneity that cannot be completely described with a limited number of field samples,
2. measurement errors, and
3. structural differences between the real-world system and the model used to represent it.

A structurally perfect model that uses inputs estimated from perfect measurements of a homogeneous real world will produce perfect predictions. Departures from this ideal situation can be attributed to one or more of the above error sources.

Once the fundamental sources of model error are identified, prediction accuracy may be investigated by analyzing the model's sensitivity to changes in appropriate error source variables. Monte Carlo simulation provides a particularly convenient method of analysis (Chu et al., 1987; Graham and McLaughlin, 1989a; Smith and Schwartz, 1980, 1981a,b). A Monte Carlo analysis based on Figure 4.1 essentially repeats the entire modeling process—sampling, input estimation, and prediction—many times. Each of these hypothetical modeling studies (or replicates) is based on a different synthetically generated real-world description and a different synthetically generated set of measurement errors. The prediction error obtained from a given replicate is simply the difference between the model prediction and the corresponding real-world value. Prediction confidence intervals and other related statistics can be readily computed from the complete ensemble of prediction error replicates. This generalization of traditional model sensitivity analysis includes field sampling and input estimation as well as the model proper. An application to ground water transport is described in detail in Chu et al. (1987).

It might be argued that Monte Carlo validation methods are too complex and time-consuming to be practical in most hazardous waste modeling applications. Although this viewpoint may be true, it should be remembered that much of the controversy surrounding the use of models ultimately stems from differing assessments of model accuracy. Typically, a judge, jury, or administrator is asked to decide between two differing conclusions, both based on ostensibly competent modeling studies. If the confidence intervals associated with the model predictions are greater than the difference between the predictions, this difference cannot be considered meaningful, at least in a statistical sense. There is no way to know whether such a situation has occurred without carrying out a serious investigation of model accuracy. It seems likely that courts, regulatory agencies, and other users of model results will press for more rigorous approaches to model validation as dependence on model results becomes more common. This is clearly the motivation behind recent proposals to establish common modeling standards and quality assurance criteria (see the sections on model quality assurance below). In the future, modelers will probably have to devote more effort to validation and related error analysis if their models are to have any credibility in public or legal forums.

ASSURING THE QUALITY OF MODELS

A successful application of a model requires knowledge of scientific principles, mathematical methods, and site characterization, paired with expert insight into the modeling process. Previous chapters discuss the types of processes that can be modeled and the formulation of models. The issues discussed are certainly complex and require a concerted effort by the user of the model. A practitioner approaching a modeling study is presented with two challenges: (1) formulating the model, including the development of boundary conditions and input parameters, and (2) the more mundane task of documenting and checking the modeling process. Most modelers enjoy the modeling process but find less satisfaction in the process of documentation and quality assurance (QA). However, both aspects are equally important to a successful application of a model. Documentation of all aspects of the modeling is important to ensure that the study would be reliably repeated.

Quality assurance is defined as the procedural and operational framework used by an organization managing the modeling study to assure technically and scientifically adequate execution of all tasks included in the study and to assure that all modeling-based analysis is reproducible and defensible (Taylor, 1985). This definition will be used in this report. In ground water modeling, QA is crucial to both development and application of the model and should be an integral part of planning, applied to all phases of the modeling process.

Adequate documentation and other forms of QA are becoming increasingly important as applications of models become part of regulatory submittals and are used to judge regulatory compliance. Both the regulators and the regulated community have an obligation to provide a complete picture of any modeling study. The section below on "Quality Assurance Procedures for Code Development" identifies some of the issues that may arise during modeling and provides some direction on documenting the modeling process. The information contained in this chapter is necessarily limited. The reader is directed to the reference list for additional information on various aspects of quality assurance.

Certain negative elements may be associated with poorly conceived and implemented QA plans. For example, there is the risk that a QA checklist will serve only to instill false confidence in model results. Another problem can be that the time and the cost of following bureaucratically imposed QA procedures may be so great that funds available for data collection and hydrogeologic analysis in any given problem could be significantly depleted.

Quality assurance procedures are not generally embraced by the modeling community. There is a perception that it has not been convincingly demonstrated that QA programs improve the quality of models. Some individuals believe that if a model is developed by qualified people who understand both the processes being simulated and the properties and boundaries of the area being represented, then it is very unlikely that QA would yield a better model (and if QA costs are high, QA may actually yield a worse model). Further, if unqualified or unprepared people are doing the work, then it is unclear that a formal QA procedure would yield a better model. (It may, however, yield unwarranted confidence in the model because a QA procedure was followed.)

The committee believes in the need for QA procedures as a key element in contemporary modeling studies. The important benefits of well-designed QA procedures need to be highlighted and demonstrated through programs of applied research. QA concepts will make a positive contribution to the way models are constructed and used; however, there is certainly a need to refine and extend existing approaches.

QUALITY ASSURANCE PROCEDURES FOR CODE DEVELOPMENT

The most important QA procedures in code development and maintenance applicable to ground water models are (van der Heijde, 1987) as follows:

- verification of structure and coding;
- validation of theoretical basis (model validation);
- documentation of code development and testing (recordkeeping);
- documentation of characteristics, capabilities, and use of code (software documentation); and
- scientific and technical reviews.

If any modifications are made to the model coding for a specific problem, the code should be tested again; all QA procedures for model development should again be applied, including accurate recordkeeping and reporting. All new input and output files should be saved for inspection and possible reuse. The following subsections briefly describe the various elements of QA procedures.

Verification of Program Structure and Code

The objective of the code verification process is twofold: (1) to demonstrate that the computational algorithms can accurately solve the governing equations and (2) to assure that the computer code is fully operational.

To check the code for correct coding of theoretical principles, for code logic, and for major programming errors ("bugs"), the code is run with specially designed problems. The computational algorithms embedded in the code are often tested using problems for which an analytical solution exists. This stage of code testing is also used to evaluate the sensitivity of the code to the design of the grid, to various dominant processes, and to a wide selection of parameter values (Gupta et al., 1984; Huyakorn et al., 1984).

Although testing numerical computer codes by comparing results for simplified situations with those of analytical models does not guarantee a fully debugged code, a well-selected set of problems ensures that the main program and most of the subroutines are used in the testing. The effectiveness of a verification exercise can be further enhanced by using a so-called walk-through, that is, a step-by-step analysis of the program operation using the data from the test cases.

A major problem with testing numerical computer codes is that analytical models are available only for simple flow or mass transport problems. The situations that numerical models are built to deal with (e.g., heterogeneous system properties and irregular boundaries) cannot be evaluated. In effect, it often has to be assumed that because the model is accurate for simple problems it will be accurate for complex conditions. Therefore, as part of the verification process, hypothetical problems might be used to test special computational features that are not represented in simple, analytical models, as in testing for irregular boundaries, varying boundary conditions, or certain heterogeneous and anisotropic aquifer properties. These hypothetical problems can be simulated by independent codes and the results compared. Model developers are in the best position to produce a comprehensive set of verification tests for their models because they are most familiar with the structure of the coding.

Model Validation

The objective of model validation is to determine how well the mathematical representation of the processes describes the actual system behavior in terms of the degree of correlation between model calculations and actual measured data. Ideally, results should be compared to the results of a well-defined field experiment or a well-conditioned laboratory experiment.

Validation of the predictive capabilities of the model is accomplished through comparison with experimental data by using independent estimates of the parameters. In principle, this is the ideal approach to validation. However, unavailability and inaccuracy of field data often prevent the application of such a rigid validation approach to actual field systems. Typically, parts of the field data are designated as calibration data, and a calibrated site model is obtained through reasonable adjustment of parameter values. Other parts of the field data are designated as validation data; the calibrated site model is used in a predictive model to simulate similar data for comparison. Although this procedure will not allow complete validation of a modeling process, it will provide some insight into potential problems in model use. This approach is limited because the splitting of the ground water data set into two components (to be used for two purposes) does not yield two independent sets of data. Two independent sets of data do not occur because of the slowness of responses in ground water systems and the production of persistence in memory over time, so that the calibration data and validation data are related.

Model validation is, of course, supposed to yield a valid model. However, a valid model is an unattainable goal of model validation. Several different groups in ground water modeling have in the past defined and used the terms validate and verify to mean different things. In fact, there is no consensus among ground water hydrologists, either on the definition of these two terms or on how to achieve validation and verification.

These terms and their underlying concepts and implications are among the contributing factors that have led to some abuses and misuses of models. Some people may feel that once a model has been designated as being validated, it does not require further questioning or testing. They may then apply the model to make predictions for conditions in which stresses lie outside their historically observed range (or outside the range of the calibration data) and produce an unreliable prediction (or an invalid prediction from a supposedly valid model).

Conversely, do uncalibrated (or unvalidated or unverified) models have any value? Definitely. Deterministic models can be used to gain insight into ground water processes that may be controlling responses in a given area, even if too few data are available from that area to yield a satisfactory calibration. Such preliminary models can help the analysts improve their understanding of the problem, formulate hypotheses to be tested, and prioritize data collection efforts.

Absolute validity of a model is never determined. Establishing absolute validity requires testing over the full range of conditions for which the model is designed, an exercise that is often not possible or practical. For many types of models this is due to the lack of adequate, high-quality field data. Thus testing of ground water models is generally limited to extended verifications, using existing analytical solutions, and to code comparisons.

In the comparison of codes, a newly developed model is compared with established models designed to solve the same type of problems. If the results from the new code do not deviate significantly from those obtained with the existing code, a relative or comparative validity is established. However, if significant differences occur, in-depth analysis of the results and codes is required. If code comparison was used to evaluate a new code, all the involved models should again be validated as soon as adequate data sets become available.

Various approaches to field validation of a model are viable. Therefore the validation process should start with defining validation scenarios. Field validation should include the following steps (Hern et al., 1985):

- Define data needs for validation and select an available data set or arrange for a site.
- Assess the quality of data in terms of accuracy (measurement errors), precision, and completeness.
- Define performance or acceptance criteria of the model.
- Develop strategy for analysis of sensitivity.
- Perform validation runs and compare performance of the model with established acceptance criteria.
- Document the validation exercise in detail.

Recordkeeping

Quality assurance for development and maintenance of codes should include complete recordkeeping of the development, modifications, and phase validation of the code. The paper trail for QA in the development phase consists of reports and files on the development and testing of the model.

Software Documentation

Software documentation explains all pertinent aspects of the system represented in the software, including purposes, methods, logic, relationships, capabilities, and limitations (Gass, 1979). Complete documentation consists of information recorded during the design, development, and maintenance of computer applications. It is the principal instrument used by those involved in a modeling effort, such as authors, programmers, users, and system operators, to communicate efficiently regarding all aspects of the software.

Good documentation includes a complete treatment of the equations on which the model is based, the underlying assumptions, the boundary conditions that can be incorporated in the model, the method used to solve the equations, and any limitations related to the particular method of solution. The documentation must also include instructions for operating the code and preparing data files, example problems complete with input and output, programmer's instructions, operator's instructions, and a report of the verification. The importance of clear documentation cannot be overemphasized. Improper documentation will prevent a code from being adequately reviewed and could propagate errors in code use. Documentation should commence at the very beginning of a software development project.

Scientific and Technical Reviews

Generally, the complete scientific and technical review process is qualitative in nature and comprises examination of model concepts, governing equations, and algorithms chosen, as well as evaluation of documentation and general ease of use, inspection of the structure of the program and the logic, handling of errors, and examination of the coding (Bryant and Wilburn, 1987; van der Heijde et al., 1985b). If verification or validation runs have been made, the review process should include evaluation of these processes.

To facilitate thorough review of the model, detailed documentation of the model and its developmental history is required, as is the availability of the source code for inspection. In addition, to ensure independent evaluation of the reproducibility of the results of verification and validation, the computer code should be available for use by the reviewer, together with files containing the original test data used in the verification and validation.

MODEL APPLICATION

Quality assurance in model application studies includes review of the selection of data, data analysis procedures, methodology of modeling, and administrative procedures and auditing. To a large extent, the quality of a modeling study is determined by the expertise of the team involved in the modeling and quality assessment.

In many cases, the person developing the model may never have seen or visited the field area. This can easily lead to fatal flaws in the model design and parameter estimation, because significant (and perhaps obvious) hydrogeologic features are not recognized and incorporated into the model. This problem may be exacerbated if the model designer or user has expertise with mathematics, numerical analysis, and/or computer simulation methods, but little field experience. In larger organizations, it is common to have such a division of labor, particularly when projects have relatively short deadlines. Where efforts are so divided, modeling may be performed by an entirely different and separate group of specialists. Unfortunately, this may produce a tendency for the model to become an end unto itself, rather than a means (or one of many tools) for analysis and problem solving. There may also be a tendency for such projects to not fully recognize or accommodate the need for the model developer to have familiarity with the field area or allow time for analysts to benefit from feedback between the model analysis and the field investigation. On the other hand, if the same person or team of analysts is performing data analysis, data collection, and modeling, it is more likely that the model will include realistic and appropriate boundary conditions, system properties, and discretization. It is too easy to calibrate (validate) a model while being unaware of major springs, pumping wells, and surface drains or ditches (or other features) that may be controlling ground water levels and gradients in an area. Ignorance of one feature is compensated by errors in values specified for other parameters. A locally steep hydraulic gradient that exists because of a drainage ditch may be interpreted as indicating a low-transmissivity zone. Such ignorance of the field area can lead to a model that matches historical data but fails in a predictive mode. (Prediction, of course, is purportedly one of the primary values and incentives for using deterministic models. If the goal were merely to achieve a best fit to observed data, then a purely statistical model, such as a multiple regression equation, would most easily meet that objective.) Thus a well-calibrated and validated model is not necessarily an accurate or a reliable one. This fact is supported by Freyberg (1988), who reports on a numerical experiment in which nine groups of analysts used the same numerical model and identical sets of observed data to calibrate the model and predict the response to a specified change in a boundary condition. Success in prediction was unrelated to success in matching observed heads, and good calibration alone did not lead to good prediction.

In summary, a ground water model (or any scientific model or theory, for that matter) can never be proven, verified, or validated in the strictest sense of the terms by agreement with a specific set of observations. Rather, a model can only be invalidated by disagreement with observations. Agreement should serve only to increase confidence in the theory or model.

Quality assurance in code application should cover all facets of the modeling process. It should address issues such as the following:

- project description and objectives;
- correct and clear formulation of problems to be solved;
- type of modeling approach to the project;
- conceptualization of system and processes, including hydrogeologic framework, boundary conditions, stresses, and controls;
- detailed description of assumptions and simplifications, both explicit and implicit (to be subject to critical peer review);
- data acquisition and interpretation;
- model selection or justification for choosing to develop a new model;
- model preparation (parameter selection, data entry, or reformatting, gridding);
- the validity of the parameter values used in the model application;
- protocols for estimating values of controlling parameters and for steps to be followed in calibrating a model;
- level of information in computer output (variables and parameters displayed, formats, layout);
- identification of calibration goals and evaluation of how well they have been met;
- role of sensitivity analysis;
- postsimulation analysis (including verification of reasonability of results, interpretation of results, uncertainty analysis, and the use of manual or automatic data processing techniques, as for contouring);
- establishment of appropriate performance targets (e.g., a 6-ft head error should be compared with a 20-ft head gradient or drawdown, not with the 250-ft aquifer thickness) that recognize the limits of the data;
- presentation and documentation of results; and
- evaluation of how closely the modeling results answer the questions raised by management.

In exceptional circumstances, it may be possible to conduct what has come to be referred to as a postaudit. A postaudit compares model predictions to the actual outcome in field conditions. Although postaudits are used primarily to determine the success rate of a model application, positive results of a well-executed postaudit analysis contribute to the acceptability of the model itself. To use a postaudit successfully in conceptualization, assumptions, and system parameters and stresses, it should be evaluated and, if necessary, updated and the model rerun to facilitate comparison of predictions with recent, observed system responses. The importance of postaudit studies has been outlined by Konikow (1986), Lewis and Goldstein (1982), and Person and Konikow (1986). An example that illustrates the importance of postaudits is the Snake River plain case study described in Chapter 5.

An increasing number of costly decisions are made in part on the basis of the outcome of modeling studies. In the light of major differences noted in comparative studies on model application (e.g., Freyberg, 1988; McLaughlin and Johnson, 1987), it should not come as a complete surprise that several groups modeling the same problem may obtain different results. While this is not a QA issue, provisions might have to be made to resolve the inconsistencies in the modeling effort. A third team or a panel can be created to review and compare the results of both modeling efforts and to assess the importance and nature of differences present.

Quality assurance is the responsibility of both the project team and the contracting or supervising organization. It should not drive or manage the direction of a project, nor is it intended to be an after-the-fact filing of technical data.

Although the need for QA programs is apparent, the extent to which they are being applied in practice can be variable. For example, EPA rarely uses peer review for models applied in the Superfund and Resource Compensation and Recovery Act (RCRA) programs. Only recently has EPA provided a checklist of steps that a modeler must take to assure that a model is valid. When careful peer review and oversight of the development and application of contaminant transport models have been performed, the quality of the modeling has been good (see the S-Area case study).

The application of contaminant models can be greatly improved by the use of peer review experts. Every model used by or relied on by EPA, including those in the Superfund program, should go through peer review. (Various groups have endorsed peer review in the regulatory system, e.g., Administrative Conference of the United States, Recommendation No. 82-5, Advisory Panels No. 1, 1 CFR 305.82.5, 1988.) This review could involve the mathematical code, the hydrogeological/chemical/ biological conceptualization, the adequacy of the data, and the application of the model to the site-specific data. Additionally, the peer review should consider whether the prediction being called for exceeds the scientific validity of the model, e.g., the prediction of a concentration over a 10,000-yr period with a model validated over a 10-yr period.

BIBLIOGRAPHY

Adrion, W. R., M. A. Branstad, and J. C. Cherniasky. 1981. Validation, Verification, and Testing of Computer Software. NBS Special Publication 500-75, Institute for Computer Science and Technology, National Bureau of Standards, U.S. Department of Commerce, Washington, D.C.

Adrion, W. R., M. A. Branstad, and J. C. Cherniasky. 1982. Validation, verification, and testing of computer software. ACM Computing Surveys 14(2), 159–192.

American Society for Testing and Materials. 1984. Standard Practices for Evaluating Environmental Fate Models of Chemicals. Annual Book of ASTM Standards, E 978-84, Philadelphia.

Beljin, M. S., and P. K. M. van der Heijde. 1987. Testing, Verification, and Validation of Two-Dimensional Solute Transport Models. In Groundwater Contamination: Use of Models in Decision-making, Y. Y. Haimes and J. Bear, eds. D. Reidel, Dordrecht, The Netherlands.

Boutwell, S. H., S. M. Brown, B. R. Roberts, and D. F. Atwood. 1985. Modeling Remedial Actions at Uncontrolled Hazardous Waste Sites. EPA/540/2-85/001, Environmental Protection Agency, OSWER/ORD, Washington, D.C.

Brandstetter, A., and B. E. Buxton. 1987. The role of geostatistical, sensitivity and uncertainty analysis in performance assessment. In Proceedings of the 1987 DOE/AECL Conference on Geostatistical, Sensitivity and Uncertainty Methods for Groundwater Flow and Radionuclide Transport Modeling, September 15–17. U.S. Department of Energy, Washington, D.C.

Bryant, J. L., and N. P. Wilburn. 1987. Handbook of Software Quality Assurance Techniques Applicable to the Nuclear Industry. NUREG/CR-4640, Office of Nuclear Reactor Regulation, U.S. Nuclear Regulatory Commission, Washington, D.C.

Carrera, J., and S. P. Neuman. 1986. Estimation of aquifer parameters under transient and steady state conditions, 1. Maximum likelihood method incorporating prior information. Water Resources Research 22(2), 199–210.

Chu, W., E. W. Strecker, and D. P. Lettenmaier. 1987. An evaluation of data requirements for groundwater contaminant transport modeling. Water Resources Research 23(3), 408–424.

Cochran, W. G., and G. M. Cox. 1957. Experimental Designs. Wiley, New York.

Codell, R., and S. Silling. 1984. Draft Quality Assurance Plan for Operational Software. Memorandum 3110/DC/84/09/20/0, Division of Waste Management, U.S. Nuclear Regulatory Commission, Washington, D.C.

Council of State Governments. 1985. Risk Management and the Hazardous Waste Problems in State Governments. Prepared for the National Science Foundation.

Cox, D. R. 1958. Planning of Experiments. Wiley, New York.

Dagan, G. 1984. Solute transport in heterogeneous porous formations. Journal of Fluid Mechanics 145, 151–177.

Dagan, G. 1986. Statistical theory of groundwater flow and transport: Pore to laboratory, laboratory to formation, and formation to regional scale. Water Resources Research 22(9), 120s–134s.

Delhomme, J. P. 1979. Spatial variability and uncertainty in groundwater flow parameters: A geostatistical approach. Water Resources Research 15(2), 251–256.

Environmental Protection Agency. 1986. Information on Preparing Quality Assurance Narrative Statements. RSKERL-QA-2, Robert S. Kerr, Environment Research Laboratory, Ada, Okla.

Environmental Protection Agency. 1987. Evaluation of Implementation of Risk-Based Decisionmaking in RCRA, Annotated Briefing. Program Evaluation Division, Office of Policy, Planning and Evaluation, pp. 8, 9, 12, 17, 25.

Federal Computer Performance Evaluation and Simulation Center. 1981. Computer Model Documentation Guide. NBS Special Publication 500-73, Institute for Computer Science and Technology, National Bureau of Standards, Department of Commerce, Washington, D.C.

Federov, V. V. 1972. Theory of Optimal Experiments. Academic Press, New York.

Freyberg, D. L. 1988. An exercise in ground-water model calibration and prediction. Ground Water 26(3), 350–360.

Gass, S. I. 1979. Computer Model Documentation: A Review and an Approach. NBS Special Publication 500-39, Institute for Computer Science and Technology, National Bureau of Standards, U.S. Department of Commerce, Washington, D.C.

Gelhar, L. W. 1986. Stochastic subsurface hydrology from theory to application. Water Resources Research 22(9), 135S–145S.

Gelhar, L. W., and C. L. Axness. 1983. Three dimensional stochastic analysis of macrodispersion in aquifers. Water Resources Research 19(1), 161–180.

General Accounting Office. 1987. Report to Congress: Superfund Improvements Needed in Work Force Management. GAO/RCED-88-1, pp. 2, 3.

Graham, W., and D. McLaughlin. 1989a. Stochastic analysis of nonstationary subsurface solute transport, 1. Unconditional moments. Water Resources Research 25(2), 215–232.

Graham, W., and D. McLaughlin. 1989b. Stochastic analysis of nonstationary subsurface solute transport, 2. Conditional moments. Water Resources Research, in press.

Gupta, S. K., C. R. Cole, F. W. Bond, and A. M. Monti. 1984. Finite-Element Three-Dimensional Ground-Water (FE3DGW) Flow Model: Formulation, Computer Source Listings, and User's Manual. ONWI-548, Battelle Memorial Institute, Columbus, Ohio.

Harlan, C. P., and G. F. Wilkinson. 1988. High-Level Waste Management Code Maintenance and Quality Assurance. SAND87-2254/UC-70. Sandia National Laboratories, Albuquerque, N. Mex.

Hern, S. C., S. M. Melancon, and J. E. Pollard. 1985. Generic steps in the field validation of Vadose Zone fate and transport models. Pp. 61–80 in Vadose Zone Modeling of Organic Pollutants, S. C. Hern and S. M. Melancon, eds. Lewis Publishers, Chelsea, Mich.

Hoeksema, R. J., and P. Kitanidis. 1984. An application of the geostatistical approach to the inverse problem in two-dimensional groundwater modeling. Water Resources Research 20(7), 1003–1020.

Hoffman, F. O., and R. H. Gardner. 1983. Evaluation of uncertainties in radiological assessment models. In Radiological Assessment: A Text on Environmental Dose Analysis, J. E. Till and H. R. Meyer, eds. NUREG/CR-3332, ORNL-5968, U.S. Nuclear Regulatory Commission, Washington, D.C.

Huyakorn, P. S., A. G. Kretschek, R. W. Broome, J. W. Mercer, and B. H. Lester. 1984. Testing and Validation of Models for Simulating Solute Transport in Groundwater: Development, Evaluation, and Comparison of Benchmark Techniques. GWMI 84-13, International Ground Water Modeling Center, Holcomb Research Institute, Butler University, Indianapolis, Ind.

Institute of Electrical and Electronic Engineers, Inc. 1984. Standard for Software Quality Assurance Plans. IEEE Standard 730-1984, New York.

Intera Environmental Consultants, Inc. 1983. A Proposed Approach to Uncertainty Analysis. ONWI-488, Battelle Memorial Institute, Columbus, Ohio.

International Association of Hydrological Sciences. 1986. Integrated Design of Hydrological Networks, Proceedings of the Budapest Symposium. IAHS Publication 158.

International Ground Water Modeling Center. 1986. U.S. EPA Ground-Water Modeling Policy Study Group: Report of Findings and Discussion of Selected Ground-Water Modeling Issues. Holcomb Research Institute, Butler University, Indianapolis, Ind., pp. 2, 5, 15, 19, 27.

Journel, A. G., and Ch. I. Hiujbregts. 1978. Mining Geostatistics. Academic Press, San Diego, Calif.

Kiefer, J., and J. Wolfowitz. 1959. Optimum designs in regression problems. Annals of Mathematical Statistics 30, 271–294.

Kincaid, C. T., J. R. Morrey, and J. E. Rogers. 1984. Geohydrological Models for Solute Migration, Vol. 1, Process Description and Computer Code for Selection. EA 3417.1, Electric Power Research Institute, Palo Alto, Calif.

Knopman, D. S., and C. I. Voss. 1987. Behavior of sensitivities in the one-dimensional advection-dispersion equation: Implications for parameter estimation and sampling design. Water Resources Research 23(2), 253–272.

Knopman, D. S., and C. I. Voss. 1988. Further comments on sensitivities, parameter estimation, and sampling design in one-dimensional analysis of solute transport in porous media. Water Resources Research 24(2), 225–238.

Konikow, L. F. 1986. Predictive accuracy of a ground-water model—Lessons from a post-audit. Ground Water 24(2), 677–690.

Lewis, B. D., and F. J. Goldstein. 1982. Evaluation of a Predictive Ground-Water Solute Transport Model at the Idaho National Engineering Laboratory, Idaho. Water Resources Investigation 82-25, U.S. Geological Survey, Reston, Va.

McLaughlin, D., and W. K. Johnson. 1987. Comparison of three groundwater modeling studies. ASCE Journal of Water Resources Planning and Management 113(3), 405–421.

McLaughlin, D., and E. F. Wood. 1988a. A distributed parameter approach for evaluating the accuracy of groundwater predictions, 1. Theory. Water Resources Research 24(7), 1037–1047.

McLaughlin, D., and E. F. Wood. 1988b. A distributed parameter approach for evaluating the accuracy of groundwater predictions, 2. Application to groundwater flow. Water Resources Research 24(7), 1048–1060.

Mercer, J. W., and C. R. Faust. 1981. Ground Water Modeling. National Water Well Association, Dublin, Ohio.

National Bureau of Standards. 1976. Guidelines for Documentation of Computer Programs and Automated Data Systems. FIPS 38, Federal Information Processing Standards Publications, Department of Commerce, Washington, D.C.

National Research Council. 1988. Hazardous Waste Site Management: Water Quality Issues. Report on a colloquium sponsored by the Water Science and Technology Board. National Academy Press, Washington, D.C., p. 9.

Neuman, S. P., C. L. Winter, and C. M. Newman. 1987. Stochastic theory of field-scale Fickian dispersion in anisotropic porous media. Water Resources Research 23(3), 453–466.

Nicholson, T. J., T. J. McCartin, P. A. Davis, and W. Beyeler. 1987. NRC experiences in HYDROCOIN: An international project for studying groundwater flow modeling strategies. In Proceedings, GEOVAL 87, April 4–9, Stockholm. Swedish Nuclear Power Inspectorate, Stockholm.

Office of Technology Assessment. 1982. Use of Models for Water Resources Management, Planning, and Policy. Pp. 9, 18–19, 20, 22, 24, 102–104.

Person, M., and L. F. Konikow. 1986. Recalibration and predictive reliability of a solute-transport model of an irrigated stream-aquifer system. Journal of Hydrology 87, 145–165.

Pesticide and Toxic Chemical News (Food Chemical News). 1987. Exposure estimates seen as weak link in EPA risk assessments. Vol. 16, p. 5 (December 23, 1987).

Rao, P. S. C., R. E. Jessup, and A. C. Hornsby. 1981. Simulation of nitrogen in agro-ecosystems: Criteria for model selection and use. Pp. 1–16 in Nitrogen Cycling in Ecosystems of Latin America and the Caribbean, Proceedings of the International Workshop, Cali, Colombia, March 16–21.

Silling, S. A. 1983. Final Technical Position on Documentation of Computer Codes for High-Level Waste Management. NUREG/CR-0856-F, Office of Nuclear Material Safety and Safeguards, U.S. Nuclear Regulatory Commission, Washington, D.C.

Simmons, C. S., and C. R. Cole. 1985. Guidelines for Selecting Codes for Groundwater Transport Modeling of Low-Level Waste Burial Sites, Vol. 1, Guideline Approach. PNL-4980 Vol. 1. Battelle Pacific NW Laboratory, Richland, Wash.

Smith, L., and F. W. Schwartz. 1980. Mass transport, 1. A stochastic analysis of macroscopic dispersion. Water Resources Research 16(2), 303–313.

Smith, L. and F. W. Schwartz. 1981a. Mass transport, 2. Analysis of uncertainty in predictions. Water Resources Research 17(2), 351–369.
Smith, L. and F. W. Schwartz. 1981b. Mass transport, 3. Role of hydraulic conductivity. Water Resources Research 17(5), 1463–1479.
Sykes, J. F., S. B. Pahwa, D. S. Ward, and R. B. Lantz. 1983. The validation of SWENT, a geosphere transport model. In Scientific Computing, R. Stepleman et al., eds. IMACS/North-Holland Publishing Company, New York.
Taylor, J. K. 1985. What is quality assurance? Pp. 5–11 in Quality Assurance for Environmental Measurements, J. K. Taylor and T. W. Stanley, eds. ASTM Special Technical Publication 867, American Society for Testing and Materials, Philadelphia.
van der Heijde, P. K. M. 1984. Availability and applicability of numerical models for groundwater resources management. In Practical Applications of Ground Water Models, Proceedings NWWA/IGWMC Conference, Columbus, Ohio, Aug. 15–17. National Water Well Association, Dublin, Ohio.
van der Heijde, P. K. M. 1987. Quality assurance in computer simulations of groundwater contamination. Environmental Software 2(1), 19–28.
van der Heijde, P. K. M., and M. S. Beljin. 1988. Model Assessment for Delineating Wellhead Protection Areas. EPA 440/6-88-002, Office of Ground-Water Protection, Environmental Protection Agency, Washington, D.C.
van der Heijde, P. K. M., and R. A. Park. 1986. U.S. EPA Groundwater Modeling Policy Study Group, Report of Findings and Discussion of Selected Groundwater Modeling Issues. International Ground Water Modeling Center, Holcomb Research Institute, Butler University, Indianapolis, Ind.
van der Heijde, P. K. M., Y. Bachmat, J. Bredehoeft, B. Andrews, D. Holtz, and S. Sebastian. 1985a. Ground-Water Management: The Use of Numerical Models, 2nd ed. Water Resources Monograph 5, American Geophysical Union, Washington, D.C.
van der Heijde, P. K. M., P. S. Huyakorn, and J. W. Mercer. 1985b. Testing and validation of ground water models. In Proceedings, NWWA/IGWMC Conference on Practical Applications of Groundwater Models, Columbus, Ohio, Aug. 19–20. National Water Well Association, Dublin, Ohio.
van der Heijde, P. K. M., A. I. El-Kadi, S. A. Williams, and D. L. Cave. 1988. Groundwater Modeling: An Overview. GWMI 88-10. International Ground Water Modeling Center, Holcomb Research Institute, Butler University, Indianapolis, Ind.
van Tassel, D. 1978. Program Style, Design, Efficiency, Debugging, and Testing, 2nd ed. Prentice-Hall, Englewood Cliffs, N.J.
Ward, D. S., M. Reeves, and L. E. Duda. 1984. Verification and Field Comparison of the Sandia Waste-Isolation Flow and Transport Model (SWIFT). NUREG/CR-3316, Office of Nuclear Safety and Safeguards, U.S. Nuclear Regulatory Commission, Washington, D.C.
Wilkinson, G. F., and G. E. Runkle. 1986. Quality Assurance (QA) Plan for Computer Software Supporting the U.S. Nuclear Regulatory Commission's High-Level Waste Management Program. NUREG/CR-4369, Office of Nuclear Safety and Safeguards, U.S. Nuclear Regulatory Commission, Washington, D.C.

Yeh, W. W-G. 1986. Review of parameter identification procedures in groundwater hydrology: The inverse problem. Water Resources Research 22(2), 95–108.

Yourdon, E., and L. L. Constantine. 1979. Structured Design: Fundamentals and Systems Design. Prentice-Hall, Englewood Cliffs, N.J.

7

Research Needs

INTRODUCTION

The committee has examined ground water modeling and the use of these models in regulation and litigation. Specifically, the committee was asked to answer two difficult questions: "To what extent can the current generation of ground water models accurately predict complex hydrologic and chemical phenomena?" and "Given the accuracy of these models, is it reasonable to assign liability for specific ground water contamination incidents to individual parties or make regulatory decisions based on long-term predictions?" This chapter summarizes the committee's recommendations for the direction and content of research programs necessary to improve the current state of affairs.

Two comments are in order before the recommendations are presented. First, the focus of this study has been the status of ground water models; and therefore associated areas of expertise (e.g., climatic scenarios and exposure assessment models), while mentioned, are not given the same consideration as ground water models. Hence, the recommended research, while acknowledging related fields of study, is biased toward ground water models and may not reflect a complete and balanced research program. Second, the questions presented above emphasize model accuracy; however, the committee notes that the accuracy of models should not be equated with the art of accurately applying models. Indeed, simulating the subsurface environment is a mixture of art and science, and an assessment of model accuracy is only one element in evaluating the confidence one should have in simulation results.

Identifying key or cornerstone issues relevant to a host of policy goals is essential so that limited resources can be devoted to the development of technology necessary to achieve national goals on the environment and economy. Certainly, as a nation we should maintain a leadership role in hydrogeologic studies for a variety of reasons; the application of ground water models in regulation and litigation is only one. Other reasons for maintaining leadership involve the estimation of natural resources and their availability, the evaluation of the safety of disposal of high-level and transuranic wastes in deep geologic deposits, and the understanding of potentially significant changes to our ecosystems (e.g., acid rain and CO_2 increases). In general, it is difficult to prioritize specific research requirements for each particular application, and this report does not attempt to do so. If research is needed to improve an aspect of hydrogeologic modeling for application to regulation or litigation, the committee makes no attempt to place that need in the context of other areas of study that will benefit from the research. Certainly, there are whole areas of ground water research that will be omitted, e.g., regional modeling of watersheds and river basins affected by global climate changes.

Another consideration that influences the committee's recommendations for future research is the present state of the science in subsurface hydrology. It is evolving; indeed it is on the threshold of a significant change in how the subsurface environment is interpreted. Current transport theory developments based on statistical interpretations of subsurface deposits may, in time, replace much of the deterministic theory. At issue are the characterization and simulation of dispersive phenomena. Central to this issue is the relationship between measurable quantities and parameters for flow and transport models. While these fundamental underpinnings to the models of conservative contaminant transport are being revisited, research continues to extend standard deterministic theory to better simulate a great variety of complex situations. Examples of extensions to deterministically based theory are multiphase flow phenomena, microbiological processes influencing water quality, and coupled geochemistry and transport models. Thus extensions of current models to more complex processes and greater spatial dimensionality are being made at the same time that foundational aspects of basic transport theory are being revisited.

The state of the practice does not reflect the state of the art, because of the scope of ongoing research and because of the strength with which opposing views are held and debated. The science has not come to grips with the gap between practice and art. Concern exists that until one can predict with confidence the migration of a conservative solute within a heterogeneous medium, one will not be able to convince a great many people of the veracity of reactive solute migration predictions. However, scientists must come to realize that modeling is used to avoid bad decisions as well as to make the best decision. Indeed, the evaluation of good alternatives may be uncertain to the degree that no clear best alternative exists. To the extent that existing field-scale models provide qualitative assessments of good versus bad, they are useful and appropriate. Such a rationale justifies the use of screening models to prioritize sites for further study and possible remediation. Research must be conducted to encourage greater acceptance of screening models and to ensure the proper expenditure of resources they influence. Resources also need to be devoted both to continue fundamental research and to decrease the gap between the state of the art and the state of the practice.

There is a recognized need to revise our current concept of modeling and modelers. Modeling needs to be redefined as a cost-effective way of interpreting all available data, to the extent that the interpretation provided by that modeling effort enables one to be comfortable in making a decision. Viewed in this way, modeling involves a spectrum of allied technologies that combine to provide the needed interpretation of subsurface events. In such a setting the modeling process would be viewed as a whole, and all subjective decisions affecting the modeling process are seen to contribute to an assessment of accuracy. Individuals responsible for model applications would be more appropriately described as analysts, rather than modelers, because of the spectrum of technologies to be applied and because of the subjective interpretations required.

The preceding remarks guide the scope of the committee's recommended research. The committee members, primarily ground water modelers, recognize that evaluation of modeling accuracy is a broad topic influenced heavily by subjective decisions made when climate scenarios are developed, site characterization plans are made and data are analyzed, and subsurface conceptual models are formalized.

The scope of research in the future must be broadened to formalize methods of recording subjective inputs and quantifying accuracy within the modeling process. The objective of model validation must be to quantify the accuracy of a model prediction for a particular application. In addition to a core effort to develop accuracy assessment methods, research must improve the methods available to gather and evaluate field data for site characterization, contaminant detection, and contaminant plume monitoring. The focus of a coordinated research program must be on the model process and its ability to predict, over the time frame of interest, the behavior of field-scale events.

USE OF MODELS

There is no doubt that increasingly greater scientific emphasis is being placed on the use of predictive computer models in ground water hydrology and geochemistry. Early applications of ground water models emphasized qualitative or relative evaluation of several alternatives. Models were used to better understand the potential impacts of alternative water use or disposal strategies. Water quantity rather than quality was the focus of this modeling, and relative comparisons appear to have been adequate to resolve litigation and regulation questions. With the full allocation or overallocation of ground water resources and the advent of ground water quality regulation, the attention of hydrologists has turned to quantitative analysis of water quantity and quality with emphasis placed on contaminant migration. The trend is toward analysis of the interrelationship between quality and quantity of the subsurface water resource and optimization of various pumping, storage, and remediation designs. The emphasis of most modeling efforts today is on providing an absolute rather than relative performance estimate.

Perhaps the most obvious example of this is in the area of storage and disposal of high-level nuclear wastes in geologic repositories (see, for example, Erdahl et al., 1985; Jacobs and Whatley, 1985). The Nuclear Waste Policy Act of 1982 (Public Law 97-425, 96 Stat. 2201, 42 USC 10101) specifies that the Department of Energy (DOE), the U.S. Nuclear Regulatory Commission (USNRC), and the U.S. Environmental Protection Agency (EPA) are responsible for doing the necessary preliminary work to permit the siting and construction of a geologic repository for high-level nuclear wastes in the United States. The only obvious method for predicting the rate of release, geochemical behavior, and rate of transport over a period of 100,000 yr is through computer modeling. Other approaches are possible, but they are at least as uncertain as computer modeling. For example, experiments can be conducted at elevated temperatures to accelerate reactions and thus to simulate longer periods of time, but there is no guarantee that acceleration resulting from higher temperatures will really simulate long periods of time at low temperature. Another approach is to examine geological sites and ancient archaeological relics for clues as to the behavior of certain chemical elements, but suitable situations are rare for implementing this strategy. All in all, computer modeling probably has at least as good a chance of yielding meaningful predictions as any of the other approaches.

A second example is the multitude of governmental agencies and private firms that increasingly rely on computer modeling techniques to investigate, predict, and guide the cleanup of natural waters contaminated by impurities that have escaped from landfills or from subsurface storage facilities. It appears that the two main objectives in the use of predictive modeling in this area are (1) to optimize the placement of test wells and monitoring wells and (2) to allow investigators to predict the future behavior of a plume of contamination. An obvious application would be to follow a plume of contamination in an aquifer backward in time and space in an effort to determine its original source. The general subject of contamination of ground water is discussed in some detail in a report by the National Research Council (1984) entitled *Groundwater Contamination.*

A third potential use of predictive modeling, which has not yet been widely recognized, is to determine what the natural background concentrations might have been in a region prior to any impact by man. This latter application may be particularly useful in establishing natural background concentrations of toxic metals in mineralized regions prior to the initiation of mining and milling.

There is little doubt that the current use of predictive computer models in interpreting and predicting the behavior of contaminants in ground water will continue and, in all probability, will increase. At the same time, as discussed in other chapters of this report, enough has been learned about the weaknesses of such models to justify the significant amount of skepticism that has also developed, both in the scientific community and in the regulatory arena. It is hoped that the proper mix of science and skepticism will be found and that the combination will allow the identification and use of a variety of predictive models that have been adequately tested and found to be appropriate, within acceptable limits of error, for a variety of field situations. This is truly a necessity for some situations, such as the disposal of nuclear wastes, that cannot be addressed in any other manner.

Emphasis on predictive rather than relative results has created an interest in the uncertainty of predictions. Unfortunately, uncertainty in estimates of ground water system behavior arises from several sources, some of which cannot be quantified. Indeed, there is no known truth to compare against when assessing uncertainty. This is the state of affairs despite the fact that a single conceptual picture of the subsurface environment does exist. Acknowledged sources of uncertainty are (1) ignorance of the true operative and dominant processes or reactions, (2) ignorance of true site characteristics leading to inaccurate boundary and initial conditions, (3) the inability to sample and quantify natural spatial and temporal variability, and (4) the extrapolative rather than interpolative character of predictions. The ability to quantify sample variability is complicated by the existence of measurement error, dissimilar data (e.g., sampling method, instrument, and volume), and quasi-periodic or random events. Clearly, sensitivity and uncertainty methods are unable to represent several of the known sources of uncertainty.

Recent work has heightened the awareness of the potential uncertainty in ground water model results and has led to some caution, or at least warnings, regarding the use of modeling results in the decisionmaking process. With regard to the use of deep geologic deposits for the disposal of nuclear wastes, Niederer (1988) believes that certainty is as important as safety. He suggests that the wise decision is to place waste where one has confidence in the performance of the geologic setting and not to place it where one merely hopes the performance will be safer. Niederer (1988) also believes that uncertainty in conceptual models is more disquieting than uncertainty in parameters, especially for flow models. His underlying concern is the potential dominance of uncertainty components that are not quantifiable. Confidence and credibility of ground water model applications depend on demonstrated applicability in every instance. Research must be undertaken to establish the framework necessary to demonstrate the applicability of models used in formulating or responding to regulation. The objective of such a demonstration is to ascertain the applicability of a given model through an assessment of accuracy and uncertainty for each situation or problem set of interest.

SCIENTIFIC TRENDS AND RESEARCH

Three basic objectives inform the recommendations for scientific research presented here: (1) to better understand and model individual processes and reactions, (2) to translate process-level understanding to sitewide simulation capability, and (3) to integrate the interdisciplinary technology needed to solve ground water contamination problems. While our understanding of subsurface processes and reactions has grown significantly in recent decades, something less than a predictive capability exists at this time. Indeed, where process and reaction models exist, field-scale observations of flow and transport have led to the realization that models based largely on laboratory- or caisson-scale studies do not provide a predictive capability at the field scale. It is also apparent that the understanding of models for some processes and reactions is not sufficient for predictive purposes in the face of complex, heterogeneous, and anisotropic environments. When process models become accepted, significant efforts are needed to translate the research results into an accepted field-scale technology. Assessments of model accuracy and validity at the field scale are an important aspect of this translation from science to application. Finally, interdisciplinary efforts that bring together site geologists, hydrologists, geochemists, geostatisticians, and health physicists are essential if ground water models and allied technologies are to be routinely applied to study and solve contamination problems with confidence.

Basic Understanding and Process Models

Two paths have been taken toward improving our basic understanding and developing more predictive ground water models: (1) the further development of mechanistic and deterministic models for individual processes and (2) the development of probabilistic models that recognize the inherent uncertainty in nature and in our ability to characterize and model the subsurface environment. Ultimately, both paths have a single objective: to understand basic processes and reactions and their interrelationships. Such an understanding will lead to predictive models of events at the field scale.

Physical processes that control or strongly influence contaminant migration in the subsurface remain an area of intense research. While relatively better understood than geochemical and microbiological processes, present conceptual and mathematical models of convection and dispersion do not provide accurate results or inspire confidence when applied to highly heterogeneous or otherwise complex environments. The probabilistic approach is seen as a way to account for the inherent uncertainty in both the subsurface structure and the knowledge of flow and transport processes.

Process Models

While considerable progress has been evident in developing mass transport as a practical tool, the hope of routinely using these models in practice lies somewhere in the future. One reason for this state of affairs is the limited ability of most models to account for the important transport processes in a realistic and convincing way. Nowhere is this problem as obvious as with the physical processes accounting for organic compound migration and the chemical and biological processes occurring for a variety of contaminants, where considerable effort will be expended to solve a few key problems. The following sections outline the trends of future research designed to improve our understanding of the processes and demonstrate the validity of coupled models.

Multiphase Fluid Flow and Transport Models

An obvious trend in research is to extend modeling capabilities to new classes of problems. A case in point is the commonly encountered problem of multiphase fluid flow and transport accompanied by dissolved component transport in water. Many of the most common organic contaminants are moderately to strongly hydrophobic. Examples are the chlorinated solvents, various petroleum constituents, pesticides, and PCBs. Modeling of the fate of hydrophobic compounds can be complicated because they can form a continuous nonaqueous phase, sorb to aquifer solids, and volatilize to a gas phase. Modeling the transport of hydrophobic materials will require that these complications be incorporated into a solute transport model.

When the organic compound forms a nonaqueous-phase liquid (NAPL), it creates three modeling difficulties. First, a significant accumulation of NAPL gives rise to multiphase or immiscible flow, a situation that is poorly understood mechanistically and difficult to describe mathematically. Thus modeling the movement of the NAPL, which is at least partly independent of the movement of the water, creates an added computational burden, if it can be described at all. A general lack of fluid retention characteristics and relative permeabilities for organic compounds or mixtures of organic compounds in the presence of water and air will greatly limit our ability to simulate multiphase fluid migration. Because of interest in the drainage and removal of hydrophobic contaminants, models of hysteresis in soil-fluid properties are essential in correctly simulating the wetting and drainage phenomena of both the organic compound and the water.

Second, the presence of an NAPL provides a long-term source for dissolution of contaminants to the aqueous phase. Description of the rate of dissolution requires knowledge of the presence of the NAPL and of the factors controlling its dissolution. Although it is probable that the solubilization is driven by the difference between the aqueous-phase concentration and the maximum solubility, the rate of dissolution is probably controlled by hydrodynamic aspects of mass transport and the presence of other contaminants. Even when the controlling factors are known, their inclusion into the model could increase the computational needs. Finally, modeling of NAPLs ultimately requires some field verification of NAPLs in subsurface systems. This presents numerous difficulties with regard to sampling and interpreting the field-scale environment. Bulk spills or disposals of NAPLs dominated by a single fluid (e.g., fuel oil or trichloroethene), do exist; however, many cases exist in which the NAPL is a mixture whose behavior in the environment can be quite complicated. Methods of sampling the subsurface and of preserving samples to determine the extent of contamination must acknowledge the variety of contaminants potentially present in soil and fluid samples. Due to the natural heterogeneity of subsurface environments, NAPLs often are not homogeneously present but are difficult to locate, especially because they can spread out into thin layers. Ultimately, the relationship between flow physics and natural spatial variability will have an impact on the interpretation of field-scale observations through an understanding of viscous fingering, i.e., the balance struck between continuum and channel flow phenomena.

Hydrophobic organic compounds also sorb onto or into aquifer and soil solids, especially soil organic matter and clays. Like NAPLs, sorbed materials can be a source of long-term, chronic water contamination as they are slowly desorbed. Solute transport modeling requires that the accumulation of sorbed material be accounted for and that the rate of desorption be described. In addition, realistic sorption relations are not necessarily linear (e.g., like partition coefficients), which gives rise to much more difficult mathematical and numerical solution requirements for nonlinear terms.

For NAPLs and sorbed contaminants, the coupling of their addition to the water with water-phase reactions, such as biodegradation, can create significant complications. For example, microorganisms degrading a dissolving solvent might be located a short distance away from the interface of the water and the NAPL; thus the dissolving compound is exposed to a biological reaction that consumes the contaminant, allows less contaminant to pass to the rest of the water, and creates an increased driving force for more dissolution. Reactions that can occur on a scale (e.g., micrometers to centimeters) much smaller than the model grid are among the most significant complications. The effect of including this microscale for a reaction is to introduce another spatial scale to transport models, which increases the computational intensity. Additionally, the phenomena controlling reactions (especially biological) for dissolving or desorbing contaminants are not easily described.

Third, some of the hydrophobic compounds (e.g., the chlorinated solvents) also are volatile and will partition to a gas phase. Thus if there are unconfined conditions and especially if there is gas production (e.g., with in situ bioreclamation or in situ aeration), some of the volatile contaminants can leave the aqueous and solid phases and go into the gas phase. Modeling of solute transport in such a situation must involve mass balances in the gas phase and description of the transfer rates between the gas phase and other phases. Not only do these requirements add to the computational demands, but they are not easily described with our current knowledge.

In summary, modeling that realistically includes hydrophobic components may become significantly more computationally intensive because of the need to keep track of nondissolved species, to describe transfer rates between phases, and to model on a small scale. Computationally efficient solution techniques, such as quasi-linearization, and the use of local analytical or pseudoanalytical solutions may become a key aspect of successful modeling.

Linking Geochemical and Physical Transport Models

Considerable success has been achieved in modeling the geochemistry of natural waters and in modeling the movement of ground waters. It is logical to take the next step and link an equilibrium geochemical model with a ground water transport model. An optimist would say that the product of the linkage should be a model that has the capability of predicting chemical changes in the ground water and reactions between the water and the aquifer at each point in space along the flow path. A pessimist would probably visualize such a linkage as being nothing more than the compounding of errors and uncertainties inherent in each of the two separate and still immature models. The truth, at this point in time, lies somewhere between the extremes, but perhaps closer to the pessimist's point of view. The basis for this somewhat negative evaluation is the fact that researchers in geochemistry have yet to demonstrate that any of the popular geochemical models can be fully validated against field or laboratory data. This is not the fault of the models, but instead points to a surprising lack of field and laboratory studies that are designed or are suitable for purposes of validating the theoretical models. Modelers tend to go their own way, building impressive computer codes to simulate nature, while field and laboratory workers tend to gather data that are highly relevant for many purposes, but perhaps not for validating models. The lack of validation is far less severe and pervasive in hydrology than it is in geochemistry, but it does exist. The main obstacle in hydrology may be the disparity between the simplifications that are required to write a usable computer code and the great complexities that can exist in real field situations. The most obvious example is the stratigraphic heterogeneity of many real aquifers, in contrast to the perfect homogeneity or the vastly simplified heterogeneity required for modeling. A similar obstacle will face geochemists when field-scale validation is undertaken.

Just as hydrologists use simplifying assumptions essential to the creation of a viable conceptual model, geochemists also employ simplifying assumptions. Foremost is the assumption of equilibrium thermodynamics determining the aqueous-phase composition. This single assumption influences the form of governing equations and thermochemical databases. Time dependency through dynamic or kinetic reactions is omitted, as are rate constants in the database. When time dependency is observed to be significant in field settings, both the reactions and the associated data will need to be incorporated into either established equilibrium-based codes or entirely new codes. It is apparent that kinetic reactions are important to some contamination events of interest, e.g., the leaching of fly ash and flue gas desulfurization sludge (Warren and Dudas, 1986).

Another aspect of equilibrium models of the aqueous phase is the natural assumption that the aqueous phase is in equilibrium with the solid phase (i.e., the porous medium). However, most unconsolidated media do not represent a solid phase in equilibrium with itself. The unconsolidated solid phase may be the result of physical (e.g., floods and glaciation) or geochemical processes. Thus the mineral composition of solids that make up porous media is often quite complex. With respect to equilibrium models, one observes that if the mineralogy of an unconsolidated porous medium were dissolved and solids were precipitated according to equilibrium reactions and data, one would not obtain the original mineralogy. This implies that a great deal of care must be taken to correctly conceptualize the geochemistry of ground waters. One must identify that portion of the solid phase that most strongly influences or defines the aqueous-phase speciation and concentration.

The solution to the dilemma of constructing fully linked geochemical and transport codes, and being able to trust the result, lies in part in a close interaction between the modelers and the field and laboratory personnel. We have probably reached the stage of development in modeling at which it is imperative to gather more and better empirical data to demonstrate the validity of geochemical and transport models.

The problem of how to deal with biological reactions is also of particular concern. As shown earlier, chemical modeling for solute transport often will involve a coupling of equilibrium and kinetic concepts: equilibrium concepts are used to determine which reactions are possible, while kinetic concepts are used to estimate the rate of possible reactions. Biological modeling usually is based solely on kinetics, although at least two types of materials must be modeled: the substrate (degradable compound) and the active microorganisms.

Several geochemical models that are useful for the equilibrium part of the computation are currently available (see Chapter 4), although they do not contain any type of kinetics. However, and very important, these geochemical models are quite complex and computationally demanding to solve for only equilibrium calculations. Incorporation in their present form into solute transport modeling is impractical because of the computing demand. Therefore the available geochemical models with comprehensive reactions and databases do not seem to be appropriate for solute transport modeling. Instead, simpler versions of these comprehensive codes have been developed and incorporated into transport theory. One example is the MICROQL:1 code (Westall, 1979) derived from the MINEQL code (Westall et al., 1976) and applied in the TRANQL code (Cederberg et al., 1985). Ideally, these more streamlined and efficient daughter codes could be tailored to include only the species and components of known importance to the site being studied. This also implies a tailored database specific to the species, components, and reactions of interest. Such an approach is the basis of the FASTCHEM™ package, a coupled transport and geochemistry code (Hostetler et al., 1988). Yeh and Tripathi (1989) have concluded that such a sequential iteration is the only practical approach for conducting realistic applications. Kinetic expressions have been incorporated into such a model (Kirkner et al., 1985); however, it remains to be seen if completely general kinetic expressions (e.g., mass transport to and from surfaces) can provide flexible source-sink algorithms for chemical reactions. Of course, great flexibility is needed in such a computational shell because there are so many types of chemical reactions exhibiting kinetic behavior.

Biological reactions present a major modeling challenge. One facet of the challenge is the complexity caused by the need to describe the accumulation of attached microorganisms, mass transport from the liquid to the microorganisms, and highly nonlinear reaction rates. In addition, multiple substrates (and sometimes products) and bacterial transport normally must be modeled. Finally, the modeling challenge is greatly increased when bacterial growth is sufficient to change the permeability and transmissivity of the aquifer. Then the biological reaction affects the pressure distribution and water flow paths, which in turn can affect the biological reaction. Thus biological reaction and flow characteristics may need to be modeled interactively.

Biological reactions also are challenging to model, because the scale of the reactions and the changes in reaction can be very small. For example, biodegradation of low concentrations of biodegradable organic compounds usually goes to completion in travel distances of only a few centimeters. Higher concentrations, lower biodegradability, and limitations from other materials can extend the distance over which biodegradation occurs; nonetheless, biological reactions often occur at a scale much smaller than a normal grid spacing. A significant challenge for the future is efficient incorporation of these small-scale phenomena into models that require larger grid spacings.

Nonuniform grid spacing and local analytical or pseudoanalytical solutions appear to be good approaches.

Before biological and chemical models can be used routinely in solute transport, the fundamental mechanisms must be studied and better understood, and the models must be tested in controlled field studies. Both steps are difficult and expensive.

Probabilistic Methods

Methods available in ground water hydrology for obtaining estimates of uncertainty involve two general approaches: deterministic porous-media models, where the probability component enters primarily through parameter variations, and stochastic porous-media models, where the probability component enters through the treatment of the medium itself as well as through parameter variations (Gutjahr, 1988). These two approaches differ in how probabilities are assigned and incorporated and how the process is modeled, i.e., as deterministic or stochastic. In the latter case, randomness is viewed as an inherent feature of flow and dispersion. That is, the properties of the media are viewed as random processes in space, and stochastic models yield results having stochastic properties. There is a single characterization of the subsurface, and acceptance and application of random field interpretations of the subsurface are a departure from this truth (Gutjahr, 1988).

Data requirements for probabilistic analyses include generic and site-specific values. For example, it is generally accepted that transmissivity has a generic log normal distribution for any particular type of material (Freeze, 1975; Freeze and Cherry, 1979; Hoeksema and Kitanidis, 1985); however, this refers to the actual values and not to the mean transmissivity. Such generic models lend a structure to the inherent randomness. Nevertheless, although the form of the distribution is assumed, sampling a highly variable system may still require that a significant quantity of site-specific data be available before the generic distribution can be applied to a site. Methods such as kriging, either standard or nonparametric, are commonly proposed and employed to analyze the spatially distributed yet relatively sparse data. Measurement error can also be accommodated by these kriging methods. In general, joint probability distribution information (e.g., interrelating transmissivities, porosities, and retardation factors) does not exist and must be created by subjectively estimating the character of correlation between variables. In addition, while stochastic models incorporate uncertainty much more directly, they also require specification of the covariance function. This involves both the covariance type and its scale parameters.

An important aspect of probabilistic methods is the performance measure used to assess the accuracy of a model and hence build confidence in the modeling process. The same performance measure should be used for calibrating, validating, and applying the model. Performance and accuracy measures for flow and transport codes should be based on their eventual applications. For example, to predict or extrapolate transport, one needs confidence in the flow model's predicted hydraulic gradient, which directly influences velocity; one does not require great confidence in the hydraulic head even though it is more readily measured. Velocity and contaminant flux should be considered performance measures for transport codes. Continued use of measures of model performance that do not reflect the modeling objective will result in an "achievable validation," which may equate to "plausible deniability" in a legal or regulatory setting.

Research is ongoing and should continue to develop both probabilistic approaches to modeling ground water systems. Both approaches require significantly larger computational resources and substantially more field data than purely deterministic models were thought to require a decade ago. Despite recent advancements, neither the stochastic nor the deterministic method has satisfactorily resolved the role of scale-dependent dispersion. With one or both of these probabilistic approaches, a broadly based technology is required to quantify uncertainty throughout the modeling process for subsurface systems, and a significant effort is needed to demonstrate the relevance of conceptual models, mathematical approaches, and characterization techniques. A holistic methodology is needed to quantify the uncertainty and relate it to measurable and meaningful field parameters.

Translation of State of the Art to State of the Practice

Research to better understand individual processes or reactions is often conducted in idealized settings that allow one to isolate cause-effect relationships; however, these settings are far from representative of field settings. Substantial efforts are required to create modeling capabilities applicable at the field scale and to demonstrate their relevance and accuracy through calibration and validation exercises. Models used in formulating or responding to regulation must be shown to be applicable and therefore valid. Essentially, accuracy should be established before application of the model to a particular site. These efforts are referred to as efforts required to translate the process-level understanding to field-scale simulation capability. Areas of research that further this translation process include field-scale code developments, validation or accuracy estimation methods, advances in computer hardware and numerical methods, and artificial intelligence or expert systems.

Field-Scale Code Developments

Several questions are asked about field-scale codes in the legal and regulatory settings. Do they embody specific state-of-the-art process models? Do they include alternative models enabling the study of opposing views? Are they valid for the proposed application? The formalism of answering and documenting the answers to these and many more questions is an aspect of quality assurance (QA) for codes. This topic is treated in Chapter 6. Whether considered research or not, a substantial resource commitment is required to create and maintain quality software.

Apart from QA issues, the development of codes applicable at the field scale is not a trivial undertaking. Flow and transport theory, which is based wholly on laboratory experience, will not necessarily apply at the field scale. Perfectly packed soil columns and highly controlled laboratory experiments seek to eliminate confounding effects of competitive processes and enable scientists to better understand individual processes or reactions or specific combinations of processes and reactions. Such controlled and contrived situations often fail to represent field-scale events. Current research on probabilistic interpretations of the subsurface environment is an attempt to quantify the uncertainty in system response arising from quantifiable and unquantifiable spatial variability in the environment.

Because of the level of current activity in field-scale modeling and the strength with which opposing views are held, it is not clear that a single state of the art can be agreed upon for conservative solute transport. Codes that embody state-of-the-art process models are relatively few in number. Because alternative models are so varied in their mathematical and computational structure, virtually no single code embodies alternative models enabling the study of opposing views. Major field-scale experiments conducted in the recent past (Betson et al., 1985; Mackay et al., 1986; Thurman et al., 1986) have not provided data sufficient to discriminate between alternative theories. Thus, the question of validity, especially for extrapolations to predict future events, remains unanswered. Some scientists who have played a dominant role in developing probabilistic theory now advocate the use of such models only by qualified professionals (Freeze et al., 1989). Probabilistic methods are sufficiently more complex than accepted single-valued deterministic models that caution should be used by the uninitiated, because model results may be instrumental in decisions influencing large populations and significant resource commitments.

Validation or Accuracy Estimation Methods

The validation of a site model, equated here to an assessment of accuracy, requires an acknowledgment of the origins of uncertainty in models of the subsurface. Characterization of actual sites is always uncertain, i.e., incomplete, and consequently, the next data point sampled may reveal a new feature of the site and dramatically alter the established conceptual model. In a very real sense, the truth about a site is never known, and hence absolute comparisons and statements can never be made. What can be addressed are the uncertainty in measured parameters and the influence of that uncertainty on the simulation of system behavior. Those aspects of a site that cannot be quantified and simulated with a single model must be addressed by ad hoc simulations of equally probable interpretations of the site, i.e., alternative conceptual models.

Calibration and validation are areas of research in ground water modeling. Site-specific data on initial and boundary conditions, along with material and fluid properties, when combined with computer software or codes, form a model of the site. A calibrated model must incorporate or explain all observations of a site. Through calibration and validation, we strive to completely understand the geology, hydrology, and geochemistry of a site. Throughout this process, it is important to acknowledge the relationships between scales of observation and modeling. Clearly, when the spatial and temporal scales of observation and the model do not match, one must interpret observations to match model scales before making comparisons. Calibration is undertaken in two ways: automated as an inverse or parameter identification method or ad hoc as a trial-and-error method.

Both approaches benefit from advances made in the last decade in the interpretation of field data. Kriging methods enable us to obtain best estimates of interpolated variables (e.g., hydraulic conductivity and hydraulic head), and at the same time they enable us to quantify estimation error. While couched in assumptions that characterize the connectivity of the physical environment, these methods have significantly broadened the view of what is possible in error and accuracy assessment. Recently, research has been directed toward nonparametric methods that will enable the blending of soft (qualitative) and hard (quantitative) data. Such methods will also make it possible to blend dissimilar data derived from different measurement methods or sampling techniques.

Advances in Computer Hardware and Numerical Methods

Current supercomputers and array processors embody an architecture that facilitates vector-based processing, and future architectures embody multiple central processing unit (CPU) designs. The availability of hardware is an important influence on future model developments. Supercomputers are becoming more widely available throughout government laboratories and universities. Of perhaps greater importance are the powerful desktop systems that are now becoming available. The computational power of these very affordable systems rivals that of many moderately sized mainframes. The speed and memory capabilities of new systems, when combined with multigrid, conjugate gradient, and moving front techniques, have enabled researchers to solve very high resolution problems. Cases requiring a million or more nodes are within reach for both two- and three-dimensional problems; however, only physical processes such as convection and dispersion have been simulated.

Hydrogeologists have always been quick to take new ideas and put them to work in models, and this trend is expected to continue as new theoretical ideas develop and new computer hardware becomes available. There are two potential areas that could yield significant scientific returns in modeling. First, the effort to construct more meaningful models, especially those that involve a linkage between geochemistry and hydrologic flow, seems to result in relentlessly increasing needs for greater computing resources. In other words, if the model is large and difficult to handle, the response of many workers in the field seems to be to "get a bigger computer." It may now be appropriate to reevaluate the approach to the need for more realistic models. Instead of simply buying bigger and better computers to handle the new codes, perhaps new mathematical methods for improving the computational efficiency of existing codes should be considered. One possible breakthrough in this area has been published by Meintjes and Morgan (1985) and Morgan (1987), in which they develop a methodology for optimizing the numerical solution of sets of simultaneous equations developed to describe chemical reactions in moving fluids; at present, the method is limited to small sets of equations, but additional mathematical research might make it useful for larger systems. As an aside, mathematical research might be more profitably conducted by individuals and by organizations that are not as richly endowed in computing resources as some of the national centers.

Second, another possible way to back away from the need for increased computer power might be to limit the magnitude of the problems to be studied and, with such a limitation, to reduce the overhead of computing resources that must be carried. For example, the databases that are part of very large geochemical models might be selectively reduced in size to be more appropriate for specific problems. Examples could include the removal of thermodynamic data for nonessential radionuclides when modeling the chemistry and movement of plumes of dissolved materials from landfills or the cooling ponds of conventional power plants.

Artificial Intelligence and Expert Systems

Another important theoretical push in modeling could come from the field of artificial intelligence, more specifically, expert systems. Expert systems are an emerging technology that could have a significant impact on ground water modeling methodologies as they exist at the present time. In essence, an expert system is a computer program that attempts to capture expert knowledge in a consistent and organized way in order to solve real-world problems. Commercial and prototype systems have been developed for many different kinds of applications. Earth science applications so far have included a variety of different aspects of geographic information systems (e.g., map design, terrain feature extraction management of geographic databases, and geographic decision support (Robinson and Frank, 1987)), interpretation of data from geophysical logging (Bonnet and

Dahan, 1983; Smith and Baker, 1983), correlation of lithologic data between boreholes (Rehak et al., 1985), advice in screening areas for potential ore deposits (Duda et al., 1979) or organic chemicals as potential ground water contaminants (Ludvigsen et al., 1986), and advice in estimating parameters for contaminant transport models (McClymont and Schwartz, 1987).

Different approaches can be used to solve problems. For example, the earlier chapters stress an approach to predicting contaminant distributions based on formal reasoning methods in a general mathematical framework. Expert system approaches look at problems using knowledge-based techniques of reasoning (Hayes-Roth et al., 1983). Thus the key ingredient of an expert system is a knowledge base, which contains judgments, rules of thumb, intuition, and advice about the problem at hand. A further requirement for an expert system (Fenves, 1986) is facilities for manipulating the knowledge base (e.g., displaying, searching, and modifying) and controlling the knowledge base (e.g., extracting information for use). This latter element of an expert system is sometimes known as the inference engine. Expert systems are also distinguished by features such as the following:

- a significant capability for interacting with the user,
- user-friendly characteristics that make the operation of the expert system and the computer transparent to the user, and
- facilities to provide advice, answer questions, and justify conclusions (Fenves, 1986).

It is not the purpose here to describe the details of expert systems. Information on the different ways of structuring knowledge within a knowledge base, building a system from scratch or with various software tools, and testing a system is provided in introductory textbooks such as those by Goodall (1985), Waterman (1986), and Weiss and Kulikowski (1984). What follows is a discussion of applications relevant to the problem of contaminant transport modeling and a view of how expert systems could be used in the future.

Some Existing Applications

A few systems are available that illustrate how expert systems could be used in contamination-related problems. However, all of these systems are essentially prototypes that have just begun to exploit the potential of this important tool. The discussion here will focus on the application of an expert system as an "intelligent assistant" in hazard evaluation and in selecting parameters for use in contaminant transport models.

An example of how an expert system can be used in hazard evaluation is provided by the prototype system DEMOTOX (Ludvigsen et al., 1986). This system is designed to help evaluate the potential for ground water contamination caused by a variety of organic contaminants added to a soil. It works by ranking contaminants using a mobility and degradation index. The index is the ratio between the time required for a contaminant front to travel through a soil zone and the half-life for biodegradation—small values indicating a greater contamination potential (Ludvigsen et al., 1986).

Ranking is a function of not only the mobility ratio but also confidence factors that reflect the quality of the available data and expert system estimations, as well as additional confidence constraints supplied by the user. This confidence factor, ranging between 0 and 1, is multiplied by the mobility and degradation index to find the final value for classification. Thus in the absence of hard information about a particular contaminant, it is conservatively ranked as a bigger threat to contaminate ground water.

Although currently a prototype, this system contains 200 rules, more than 250 facts, and numerous explanations. It is constructed using expert system tool M.1.

A second example in the application of expert systems for hazard evaluation is provided by the work of Law et al. (1986). The specific application is the determination of ground water flow directions and the permeability of units at a site. The knowledge base for this prototype system is a series of if/then production rules and an external function that establishes ground water direction as the solution to a three-point problem.

One of the most comprehensive computer systems related to the problem of ground water contamination is Expert ROKEY (McClymont and Schwartz, 1987). The main components of this system are a contaminant transport model, two expert systems (EXPAR and EXINS), and a plotting package. The transport model, which is an analytical solution from Domenico and Robbins (1985), describes the transient, three-dimensional spread of a dissolved contaminant in a unidirectional flow system. The model accounts for advection, dispersion, sorption, first-order decay, and time-varying loading of the source. The EXPAR system helps users prepare a set of input data for the model. This operation is coordinated by a set of computer forms that serves as the main user/system interface. Each form can accept parameters supplied by the user or developed with one of the family of expert systems. These expert systems access production rules, assistance programs (conventional programs that calculate parameters, e.g., hydraulic conductivity from grain size), and appropriate case studies. Like many expert systems, they explain the meaning of questions and why the questions were asked, provide general tutorial information about a process (e.g., dispersion), and check the overall validity of derived parameters within the context of a problem.

The EXINS system is a demonstration prototype to help plan a monitoring strategy for a first-stage field investigation. The strategy is based on (1) data or information used previously in EXPAR, (2) the results from the transport simulation, and (3) specific responses to questions posed to the user.

The entire package was designed for users with minimal expertise in the use of a computer and modeling. As such, it represents one of the important potential uses of expert systems in relation to contaminant transport modeling. McClymont and Schwartz (1987) present a detailed discussion of the method and its application to a practical problem.

Expert Systems in the Future

The systems applications considered so far are conventional in the way they use expert systems. In the future, work with expert systems can be expected to go beyond these applications to look at more fundamental scientific problems. One possibility is the use of expert systems as a tool for modeling in what Beck (1987) refers to as "macroscopic logical reasoning." His idea is that for very complex and poorly characterized systems, formal mathematical descriptions using differential equations may not be the most logical way of representing the system. According to Beck (1987), expert systems might be more useful in cases where

> the system's dynamics are highly non-linear . . ., a theory is in its initial phase of development (e.g., as a verbal conceptual model; crude order must be imposed on a confused and conflicting welter of experimental observations; and decision making must be conducted in a setting where a pragmatic, universal shortcut to interdisciplinary communication is a priority.

According to Hardt (1986), there is a great deal of similarity between building a simulation model and building a knowledge base. Looking at what constitutes the description of a natural problem,

Hardt (1986) finds first that the classical approach to model formulation is "rich with mathematical intuition," and that in formulating and solving the mathematical problem and in interpreting the results, "reality" is being expressed in one, not particularly unique, way. Thus a mathematical model evolves as a scientist combines "native common sense" with the thinking tools provided by mathematics. Hardt (1986) maintains that problems can also be formulated and solved very efficiently using cognitive approaches. This nonmathematical procedure involves replacing mathematical equations with qualitative equations. Typically, this approach loses some of the detail of the problem and is less quantitative (Hardt, 1986). However, if we return to Beck's (1987) message, it may be foolish to think that for poorly described systems the mathematical approach is qualitative.

The last area where expert systems may have a role to play is the area of automatic programming. Barstow (1983) examines the potential for constructing software for solving nontrivial problems in the quantitative interpretation of geophysical logs. His prototype model was designed to assist users who were not familiar with "traditional computer interfaces" in preparing mathematical models for log interpretation, in a matter of minutes. It is beyond the scope of this examination to present an overview of this complex topic, but indications from Barstow's (1983) work are that the impact of an automatic programming system can be quite dramatic. The procedure effectively removes the "programming bottleneck between conceptualization and feedback."

The feasibility of using these automatic techniques in areas of contaminant transport modeling is an application that is open for investigation. It is a tantalizing idea to be able to sit down at a keyboard and in a matter of minutes generate a prototype model. This last potential application illustrates that the potential of expert systems in hydrogeology is limited only by our imagination.

Interdisciplinary Efforts

There is a recognized need to revise our concept of modeling and modelers. A tendency exists to describe ground water models, the application of those models, and the necessary research as a logical progression and, thereby, leave the impression that modelers believe that another generation of models is the answer. In reality, this would perpetuate a myth. Although a logical step in model research is the development of standard characterizations of accuracy and methods that provide an assessment of accuracy, it is recognized that evaluation and improvement of model accuracy are not sufficient. The broader issue to be dealt with is needed research into the art of applying models to accurately simulate the subsurface; the accuracy of a ground water model is only one component of this broader topic.

There is a need to provide through integrated efforts the interdisciplinary technologies necessary to discover, characterize, and solve ground water contamination problems. These allied technologies include measurement technologies needed to characterize sites, initially detect releases, and continuously monitor plume advance. They also include optimization techniques applied to guide sample network design (i.e., location and frequency) and to guide water management practices (e.g., safe withdrawal, minimized cost of supply, or minimized cost of containing and treating). Remediation methods and models of their potential effectiveness require an interdisciplinary effort (e.g., in situ treatment using bioremediation methods will be successful only if predicated on a knowledge of contaminant location, i.e., physical behavior) and chemical and biological process models known to accurately describe subsurface response. Risk assessment methodologies, which integrate the probability of occurrence with the consequence of occurrence, are yet another area of interdisciplinary research. Currently, risk assessment methods applied to ground water systems are simply based and are used for screening or scoping newly discovered problems; however, more sophisticated modeling capabilities are needed for remediation studies. Essentially, the decisionmakers are more interested in a quantification of risk than in a quantification of contamination level.

The measurement and interpretation of pollution events offer significant challenges to measurement technology. Instrumentation has evolved to measure more accurately the behavior and effects of contaminants. This is a result of increasingly available chemical information and recent advances in material science. The use of fiber optics to transmit signals from sensors is one area of improved instrumentation. Standard and nonparametric kriging methods are being used to develop designs for sampling the environment and interpreting field data. Optimization techniques are being used to determine the location and frequency of sampling; however, it must be appreciated that the design is optimum only with respect to sampling objectives, i.e., detection, monitoring, or characterization.

One can observe the environment to detect a contaminant, monitor a known pollutant, or characterize a site. The design of sampling systems depends significantly on the objective. Thus sampling strategies must be based on interdisciplinary knowledge. Detection systems should be designed to surround or possibly underlie disposal systems. They might be tailored to only pick up a fingerprint tracer signaling first release. Therefore substantive sampling to determine the character of the contaminant plume may not be associated with detection sampling. Monitoring of a known plume will have other objectives such as observing the peak concentrations, locating the center of mass of the migrating plume, and determining the mass flux of the contaminant across significant planes in the environment (e.g., zones of capture or remediation and property boundaries). Certainly, monitoring can imply sampling for a broad number of constituents and sampling over an ever-expanding region of contamination. Finally, characterization of a site has the objective of defining media-fluid properties, boundary conditions, and initial conditions that govern the flow of water and contaminant migration at the site. One aspect of characterization is sampling to independently define process model parameters. The identification of necessary data, suitable instrumentation, appropriate sample network design, and data interpretation methods is an interdisciplinary effort. Research should proceed toward highly integrated interdisciplinary methods in order to sample the subsurface environment.

In situ remediation of contaminated soils and aquifers is a major interdisciplinary activity. Its attractiveness stems from the expense and/or practical problems of excavation followed by proper disposal or incineration. Bioremediation is the most widely applicable strategy, because most of the common organic pollutants and some of the inorganic pollutants (e.g., ammonium, nitrate, and sulfate) are amenable to biodegradation, as long as the proper environmental and microbial conditions are present. In situ remediation is particularly advantageous when the contaminants are poorly mobile, because the removal reaction can be close to the source of the contamination. Without in situ reactions, dissolution and flushing of the contaminants can require years to decades.

The modeling issues and problems discussed in Chapter 4 describe the hydrophobic contaminants and incorporation of chemical and biological reactions that are relevant to any modeling of in situ remediation. In addition, four issues are especially acute during in situ remediation:

1. The addition and extraction of water through wells or trenches create local nonhomogeneities of head, flow, and solute concentrations. Chemical and biological reactions are likely to be most intense near the nonhomogeneities. Modeling around nonhomogeneities requires, at a minimum, a tight grid spacing.

2. Flow velocities are often significantly increased in a remediation site in order to flush water and reactants through the ground. The high velocity can alter flow paths and may accentuate the effects on heterogeneities (natural or induced). Therefore modeling that includes heterogeneities is emphasized.

3. The biological and chemical reactions often will alter the permeability of the soils or aquifer, especially near the introduced nonhomogeneities. Thus models must include the interactions of flow and reaction.

4. The model must keep track of at least two reacting species: the contaminant and the added material that reacts with the contaminant. Their removals usually are linked stoichiometrically, but one or both can control the overall reaction rate. Often, many species must be followed, including products, and these species may be affected in very different manners by other mechanisms, such as sorption or volatilization.

Another area of interdisciplinary research involves the disposal of liquid hazardous wastes by subsurface injection through wells into deep aquifers. This technique began in the United States in the 1950s and 1960s and was seen as a relatively inexpensive way to prevent pollution of rivers and lakes. Depths of injection typically range from 0.25 to 1 mi below the surface (Gordon and Bloom, 1986). The liquid wastes most frequently injected into the subsurface are corrosive and reactive liquids, organics, and dissolved metals.

In 1983, EPA identified 90 facilities in the United States where 195 wells were being used for disposal of hazardous wastes (Brasier, 1986). Subsurface injection is the predominant form of hazardous waste disposal in the United States, accounting for 60 percent, or approximately 10 billion gal. In contrast, only 35 percent of hazardous wastes was disposed of in surface impoundments and 5 percent in landfills in 1981 (Gordon and Bloom, 1986). The predominance of subsurface injection as a method of disposal is largely due to the low cost in relation to other technologies. Until recently, little, if any, treatment of the wastes was required before injection. As with other

other methods of waste disposal, usable ground water has been contaminated by escaping toxic wastes from injection facilities (Gordon and Bloom, 1986).

A majority of the subsurface injection facilities are used by the chemical and petrochemical industries located in Texas, Louisiana, Ohio, Michigan, Indiana, and Illinois. All wells used for injection of hazardous materials are subject to control by the Safe Drinking Water Act (see discussion in Chapter 5) and the Resource Conservation and Recovery Act (see Chapter 5).

Prior to the initiation of injection, a vast array of chemical, physical, geological, and hydrological parameters should be considered. Chemical and physical factors include density, reactivity, viscosity, temperature, content of suspended solids, content of gases, pH, Eh, stability, and volatility. Geological and hydrological factors that should be considered include the permeability and effective porosity of the injection horizon, thickness and integrity of the aquicludes that separate the injection zone from adjacent usable aquifers, possible zones of recharge and discharge, effective porosity, content of clay and other reactivity minerals in the host formation, magnitude and direction of pressure heads, preferred paths of flow, and salinity and reactivity of indigenous water in the formation. The prospect of having to properly consider such a list of parameters prior to injection would probably cause any potential disposer to hesitate to initiate such a program.

The extreme difficulty and cost involved in obtaining adequate field and laboratory data prior to construction of deep-well injection facilities contribute to the increasing use of predictive computer modeling. Predictive modeling potentially offers a means to minimize, or at least to optimize, the drilling of numerous test and monitoring wells and possibly to fill existing gaps in knowledge. Prickett et al. (1986) discuss the application of flow, mass transport, and chemical reaction modeling to subsurface liquid injection. They point out that modeling is necessary for estimation of pressure buildup rates at the injection well and of distribution of pressure buildup in the reservoir. With regard to transport of contaminants, it would be desirable to include advection, dispersion, sorption, decay, and biochemical reaction, but at present no model can deal with the full complexity of the transport and chemical reactivity of a waste in a deep, high-pressure, high-temperature, high-salinity, subsurface environment. Prickett et al. (1986) suggest that, while it is not possible to truly simulate the transport and reactivity of injected wastes, it should be possible to model the worst-case scenario of conservative transport of all dissolved chemicals. Strycker and Collins (1987) state that additional research is needed in virtually all areas of abiotic and biotic waste interactions before definitive explanations can be given of their long-term fate.

Clearly, the deep-well injection of hazardous wastes is an area that could potentially benefit from improvements in our capabilities for modeling transport in ground water. To reach this goal, much research is needed in the coupling of transport and chemical models, so that more realistic predictions of the movement and fate of injected chemicals can be made.

POLICY TRENDS AND SUPPORT FOR RESEARCH

An EPA study found that existing ground water models do not account for all processes affecting the fate and impact of contaminants. For example, the flow and transport of organic solvents are influenced by the hysteresis in multiphase soil-fluid characteristics and by biotic and abiotic fate processes; neither is accounted for in existing and available codes. It is thought that existing models lack accuracy when confronted with a high degree of heterogeneity, and, in general, it is believed that data requirements to ensure high levels of confidence in the accuracy of predicted results are prohibitively expensive. It is disturbing to know that models lack accuracy; it is worse not to know the accuracy of the model.

Models in support of policy and in response to regulation range from generic to fully mechanistic. Generic models often require no site-specific data, embody no attenuation mechanisms, and characterize transport as a one-dimensional flow path. The need to prioritize or rank disposal sites for cleanup actions in the face of limited resources has led to the application of models requiring little or no site-specific data (Whelan et al., 1987; see vertical-horizontal spread model case study in Chapter 5). While applications of generic models will continue, it would be informative to better understand the relationship between the results of such modeling and actual site performance. For example, when generic models are used, are the worst sites always identified as being worst, and are all sites ranked in a hierarchy associated with a real risk ranking? At the other extreme, the need to assess environmental impacts from wastes previously disposed of in complex hydrogeologic systems makes it necessary to improve our understanding of complex systems. Thus complexities of process (e.g., organic compounds, dimensionality, and mathematical formulation heterogeneity, anisotropy, spatial variability, fractured media, and karst systems) must be addressed through continued research if we are to be able to realistically portray the risk of future events.

The siting regulation for new low-level waste (LLW) disposal sites (10 CFR Part 61) states that "the disposal site shall be capable of being characterized, modeled, analyzed, and monitored" (U.S. Nuclear Regulatory Commission, 1987). Thus the responsibility for being able to simulate site performance is a responsibility of the licensee. Furthermore, it is implied that hydrogeologic systems that cannot be characterized, modeled, analyzed, or monitored with confidence are to be eliminated from consideration. Thus the need to regulate LLW sites does not directly justify research on complex hydrogeologic systems. This regulation provides no guidance on measures of confidence; however, all subsurface environments are uncertain or unknown to some degree. A logical question is: What level of confidence is necessary before one can claim an ability to model or analyze a site? Methods that quantify confidence in ground water modeling results must be developed for application to any disposal site.

As models have begun to influence the assignment of liability and the assessment of long-term hazard, modeling results have begun to be viewed as quantitative rather than qualitative. Modeling results are now frequently compared to regulatory limits, and the methods used to make these comparisons are important to the proper portrayal of modeling results. Reasonable assurance is a concept that has arisen from the study of the potential for deep geologic systems to provide isolation of high-level radioactive wastes. This term refers to the interval between a realistic assessment of poor performance and a regulatory limit. It represents the interval of safety. If reasonable assurance exists that an event is safe, then it is implied that a comprehensive and defensible analysis supports the finding.

If a "bounding performance" estimate indicates good performance (i.e., does not exceed the regulatory limit), then a realistic analysis providing an estimate of mean and uncertainty ranges is unnecessary. Only in the instance depicted in Figure 7.1, when bounding performance exceeds regulatory limit, does one need to perform a realistic analysis. A realistic analysis is essentially an effort to demonstrate regulatory compliance when realistic rather than bounding models and model parameters are employed. Of course,

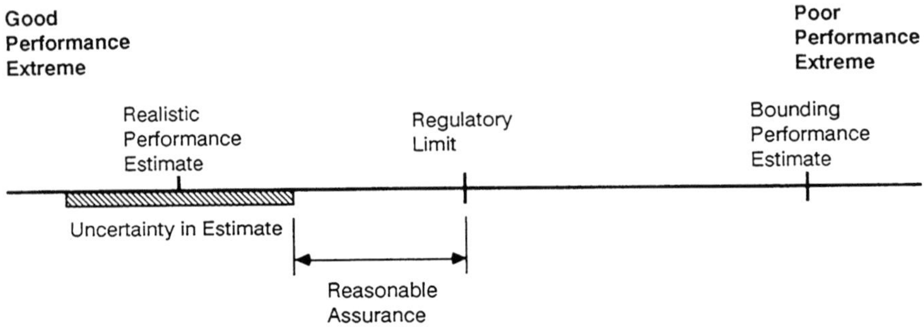

FIGURE 7.1 The relationship of reasonable assurance to bounding analysis, regulatory limit, and realistic estimates.

when realism is introduced, so is uncertainty, and it must be quantified to the extent practical. This same logic suggests that after compliance is triggered by conservative models (used in the prioritization of sites), more realism and certainty should be required if the output of a model is used directly to trigger additional regulatory action than if the model is used as an interpretative tool to better understand how contaminants migrate.

Currently, EPA is adopting an approach for pesticide regulation requiring differential management of pesticide use based on differences in the use, value, and vulnerability of ground water. This implies a recognition of the value to society of using chemicals. It may also signal movement toward acceptance of "de minimus"-based regulations, in other words, regulations based on the detection of chemicals at lower levels. Thus the ability to model complex environments and complex contaminants may become more crucial in the future.

Because model results are being viewed predominantly as quantitative in regulatory and litigious settings, accuracy and uncertainty are of interest. However, accuracy per se is difficult if not impossible to assess because the subsurface is always to some degree unknown and uncertain. Indeed, the dominant use of the term uncertainty instead of certainty implies the degree to which the environment is unknown and uncharacterizable. Current research seeks, in part, methods to quantify certainty by relating uncertainty in knowledge of the subsurface to uncertainty in predictions of future events. The "truth" of the subsurface environment is not known; therefore research toward methods of quantifying uncertainty must treat the influence of both subjective and objective judgments on model predictions.

One should be aware that in the application of an overly sophisticated model, or any model, to a situation that does not merit sophisticated modeling, the level of knowledge implied by such model results can be misleading. When mean values and/or distributions of parameters are purely assumed, assumptions may outweigh knowledge, and model results may imply a level of knowledge or certainty that does not exist. Methods of uncertainty analysis that include the influence of subjective decisions on model results will help to ensure the proper use of models by revealing cases where ignorance outweighs knowledge.

A number of governmental agencies are active in subsurface environmental studies; however, it is not clear if this contributes to the problem or to the solution of developing theoretically sound and computationally correct ground water models. For example, hydrogeologic studies are among the least funded research topics by the National Science Foundation. This is the case despite the fact that several federal agencies—including the Departments of Defense, the Interior, and Energy, as well as EPA—support a variety of research and application activities that depend on knowledge of the subsurface environment.

Issue resolution, legal or regulatory, will not wait until the perfect solution is found. The field of hydrogeology needs to have established and accepted technology, even if flawed, for application to a host of current problems while science advances. However, simply having an accepted technology does not obviate the need for continued advancement. Within the federal bureaucracy, some division exists for those who fund applications and those who perform research-oriented studies. For example, within the USNRC, the bulk of funding to support research is controlled by those responsible for licensing nuclear facilities. The foundational belief of any group having licensing responsibility must logically be that sufficiently applicable and defensible technology exists today to license needed facilities. Support for research issues requiring long-term funding and high-risk approaches may not be within their purview.

Management by crisis and/or strongly justified large initiatives appears to be the current mode of operation within government. Initiatives such as acid rain, global climate change, the supercollider, RCRA, and Superfund are examples. EPA is one of the few government agencies that have as a mission the protection and especially the improved understanding of our subsurface environment. Most have the responsibility to quantify the impact of their mission on the subsurface. Often they are charged with simply using existing technology to estimate the impacts of waste disposal, remediation, and so on, on ground water aquifers. Frequently, new initiatives encompass a spectrum of technologies, ground water environs being only one component. Acid rain and global climate change are examples of research investments that embrace ground water issues but may not significantly improve our understanding of ground water flow and contaminant transport. Rather, they will improve our understanding of linked processes that, when integrated over significant spatial and temporal scales, serve to estimate the overall response of the environment. Such diversified studies do not significantly advance our understanding of basic physical processes such as dispersion or of ways to directly relate model parameters to measurable quantities. It is true that ground water models that consider spatial and temporal changes appear to be advanced technology when compared to our understanding of geochemical and microbiological phenomena. However, more advanced methods of ground water characterization and modeling are needed in order to understand with confidence where a contaminant is in the subsurface so that the effectiveness of bioremediation methods for in situ treatment of contaminants can be estimated. Government research programs studying interdisciplinary problems need to appreciate the complexities of flow and transport phenomena that are not well understood and, as a consequence, are poorly simulated.

An interesting evolution seems to have taken place with regard to predictive modeling from the point of view of regulatory agencies. With the development of comprehensive hydrologic models in the 1960s and 1970s, regulatory agencies seemed to accept the predicted results with a certain amount of awe. The potential power of the approach was obvious to even the most nontechnical member of a regulatory board or agency. The same is true of the introduction of comprehensive geochemical models in the 1970s and 1980s. Again, the sheer power of the methodology was obvious and a bit overwhelming. Although regulatory bodies might not fully understand either the input or the output from such models, they seemed to be willing to accept the word of the experts regarding the usefulness of the predictions. However, in the last five years or so, quite the opposite attitude seems to be developing on the part of the regulatory agencies. An enormous amount of skepticism appears to have developed, with a resulting attitude of "Prove it!" having replaced the more passive and accepting faith of earlier years. At this time, modelers are in the spotlight, and on the spot, to demonstrate that their long-term predictions are worthwhile and meaningful. This new attitude can only be healthy for the science and art of predictive modeling; it will force the scientists to come to grips with the gaps and unknowns that exist, both in the models themselves and in the field and laboratory data that are required to validate the models.

REFERENCES

Barstow, D. 1983. A perspective on automatic programming. Pp. 1170–1179 in Proceedings of the Eighth International Joint Conference on Artificial Intelligence, Karlsruhe, West Germany.

Beck, M. B. 1987. Water quality modeling: A review of the analysis of uncertainty. Water Resources Research 23(8), 1393–1442.

Betson, R. P., L. W. Gelhar, J. M. Boggs, and S. C. Young. 1985. Macrodispersion Experiment (MADE): Design of a Field Experiment to Investigate Transport Processes in a Saturated Groundwater Zone. EPRI-EA-4082, Electric Power Research Institute, Palo Alto, Calif.

Bonnet, A., and C. Dahan. 1983. Oil-well data interpretation using expert system and pattern recognition technique. Pp. 185–189 in Proceedings of the Eighth International Joint Conference on Artificial Intelligence, Karlsruhe, West Germany.

Cederburg, G. A., R. L. Street, and J. O. Leckie. 1985. A groundwater mass transport and equilibrium chemistry model for multicomponent systems. Water Resources Research 21(8), 1095–1104.

Domenico, P. A., and G. A. Robbins. 1985. A new method of contaminant plume analysis. Ground Water 23(4), 476–485.

Duda, R. O., P. E. Hart, K. Konolige, and R. Reboh. 1979. A Computer-Based Consultant for Mineral Exploration. Final Report, SRI Project 6415, Artificial Intelligence Center, SRI International, Menlo Park, Calif.

Erdahl, B. R., J. H. Heiken, and J. Howard. 1985. Workshop on Fundamental Geochemistry Needs for Nuclear Waste Isolation, Los Alamos National Laboratory, N. Mex. June 20–22, 1984. Department of Energy Report CONF8406134, 208 pp.

Fenves, S. J. 1986. What is an expert system? Pp. 1–17 in Expert Systems in Civil Engineering, C. N. Kostem and M. L. Maher, eds. American Society of Civil Engineers, Seattle, Wash.

Freeze, R. A. 1975. A stochastic conceptual analysis of one-dimensional groundwater flow in non-uniform homogeneous media. Water Resources Research 11(5), 725–741.

Freeze, R. A., and J. A. Cherry. 1979. Groundwater. Prentice-Hall, Englewood Cliffs, N.J.

Freeze, R. A., G. De Marsily, L. Smith, and J. Massmann. 1989. Some Uncertainties About Uncertainty. Pp. 231–260 in Proceedings of the Conference on Geostatistical, Sensitivity, and Uncertainty Methods for Ground-Water Flow and Radionuclide Transport Modeling Held in San Francisco, California, September 15–17, 1987. Battelle Press, Columbus, Ohio.

Goodall, A. 1985. The Guide to Expert Systems. Learned Information (Europe) Ltd., Abington, England, 220 pp.

Gordon, W., and J. Bloom. 1986. Deeper problems, limits to underground injection as a hazardous waste disposal method. Pp. 3–50 in Proceedings of the International Symposium on Subsurface Injection of Liquid Wastes, March 3–5, New Orleans, La. Underground Injection Practices Council, Association of Ground Water Scientists and Engineers, Water Well Publishing Company, Dublin, Ohio.

Gutjahr, A. L. 1988. Hydrology. In Techniques for Determining Probabilities of Events and Processes Affecting the Performance of Geologic Repositories, Chapter 5. SAND86-0196, Sandia National Laboratories, Albuquerque, N. Mex.

Hardt, S. L. 1986. On the power of qualitative simulation for estimating diffusion transit times. Pp. 460–463 in Proceedings of the 1986 Winter Simulation Conference (held in Washington, D.C.), J. Wilson, J. Henriksen, and S. Roberts, eds. Association for Computing Machinery, New York.

Hayes-Roth, F., D. A. Waterman, and D. B. Lenat. 1983. An overview of expert systems. Pp. 3–29 in Building Expert Systems, F. Hayes-Roth, D. A. Waterman, and D. B. Lenat, eds. Addison-Wesley, London.

Hoeksema, R. J., and P. K. Kitanidis. 1985. Analysis of the spatial structure of properties of selected aquifers. Water Resources Research 21(4), 563–572.

Hostetler, C. J., R. L. Erikson, J. S. Fruchter, and C. T. Kincaid. 1988. Overview of the FASTCHEM^{TM} Package: Application to Chemical Transport Problems. EPRI EA-5870-CCM, Vol. 1, Electric Power Research Institute, Palo Alto, Calif.

Jacobs, G. K., and S. K. Whatley. 1985. Conference on the Application of Geochemical Models to High-Level Nuclear Waste Repository Assessment: Proceedings, Oak Ridge, Tenn., Oct. 2–5, 1984. NUREG/CP-0062, ORNL/TM-9585, U.S. Nuclear Regulatory Commission, Washington, D.C. 126 pp.

Kirkner, D. J., A. A. Jennings, and T. L. Theis. 1985. Multisolute mass transport with chemical interaction kinetics. Journal of Hydrology 76, 107–117.

Law, K. H., T. F. Zimmie, and D. R. Chapman. 1986. An expert system for inactive hazardous waste site characterization. Pp. 159–168 in Expert Systems in Civil Engineering, C. N. Kostem and M. L. Maher, eds. American Society of Civil Engineers, Seattle, Wash.

Ludvigsen, P. J., R. C. Sim, and W. J. Grenneg. 1986. A demonstration expert system to aid in assessing ground water contamination potential by organic chemicals. Pp. 687–698 in Computers in Civil Engineering, Proceedings of the Fourth Conference, W. T. Lenocker, ed. American Society of Civil Engineers, Boston, Mass.

Mackay, D. M., D. L. Freyberg, P. V. Roberts, and J. A. Cherry. 1986. A natural gradient experiment on solute transport in a sand aquifer, 1. Approach and overview of plume movement. Water Resources Research 22(13), 2017–2029.

McClymont, G. L., and F. W. Schwartz. 1987. Development and application of an expert system in contaminant hydrogeology. Unpublished report for National Hydrology Research Institute, Environment Canada, 206 pp.

Meintjes, K., and A. P. Morgan. 1985. A Methodology for Solving Chemical Equilibrium Systems. General Motors Research Laboratory Report GMR-4971, Warren, Mich., 28 pp.

Morgan, A. P. 1987. Solving Polynomial Systems Using Continuation for Engineering and Scientific Problems. Prentice-Hall, Englewood Cliffs, N.J., 546 pp.

National Research Council. 1984. Groundwater Contamination. Studies in Geophysics. National Academy Press, Washington, D.C., 179 pp.

Niederer, U. 1988. Perception of safety in waste disposal: The review of the Swiss project GEWAHR 1985. Pp. 11–26 in Proceedings of the GEOVAL 1987 Symposium in Stockholm, April 7–9, 1987. The Swedish Nuclear Power Inspectorate, Stockholm.

Prickett, T. A., D. L. Warner, and D. D. Runnells. 1986. Application of flow, mass transport, and chemical reaction modeling to subsurface liquid injection. Pp. 447–463 in Proceedings of the International Symposium on Subsurface Injection of Liquid Wastes, March 3–5, New Orleans, La. Underground Injection Practices Council, Association of Ground Water Scientists and Engineers, Water Well Publishing Company, Dublin, Ohio.

Rehak, D. R., R. R. Christiano, and D. D. Norkin. 1985. SITECHAR: An expert system component of a geotechnical site characterization work bench. Pp. 117–133 in Applications of Knowledge-Based Systems to Engineering Analysis and Design, C. L. Dym, ed. American Society of Mechanical Engineers, Miami Beach, Fla.

Robinson, V. B., and A. U. Frank. 1987. Expert systems for geographic information systems. Photogrammetric Engineering and Remote Sensing 53(10), 1435–1441.

Smith, R. G., and J. D. Baker. 1983. The dipmeter advisor system: A case study in commercial expert system development. Pp. 122–129 in Proceedings of the Eighth International Joint Conference on Artificial Intelligence, Karlsruhe, West Germany.

Strycker, A., and A. G. Collins. 1987. State-of-the-Art Report: Injection of Hazardous Wastes into Deep Wells. Report NIPER-230, National Institute of Petroleum and Energy Resources, Bartlesville, Okla., 55 pp.

Thurman, E. M., L. B. Barber, Jr., and D. LeBlanc. 1986. Movement and fate of detergents in groundwater: A field study. Journal of Contaminant Hydrology 1(1/2), 143–161.

U.S. Nuclear Regulatory Commission. 1987. Low-Level Waste Disposal Licensing Program Standard Review Plans. NUREG-1200, Washington, D.C.

Warren, C. J., and M. J. Dudas. 1986. Mobilization and Attenuation of Trace Elements in an Artificially Weathered Fly Ash. EPRI-EA-4747, Electric Power Research Institute, Palo Alto, Calif.

Waterman, D. A. 1986. A Guide to Expert Systems. Addison-Wesley, Reading, Mass., 419 pp.

Weiss, S. M., and C. A. Kulikowski. 1984. A Practical Guide to Designing Expert Systems. Rowman and Allanheld Publishers, Totowa, N.J., 174 pp.

Westall, J. C. 1979. MICROQL:1: A Chemical Equilibrium Program in BASIC, EAWAG. Swiss Federal Institute of Technology, Duebendorf, Switzerland.

Westall, J. C., J. T. Zachary, and F. M. M. Morel. 1976. MINEQL—A Computer Program for the Calculations of Chemical Equilibrium Composition of Aqueous Systems. Tech Note 18, R. M. Parsons Lab., Massachusetts Institute of Technology, Cambridge, 91 pp.

Whelan, G., D. L. Strenge, J. G. Droppo, Jr., B. L. Steelman, and J. W. Buck. 1987. The Remedial Action Priority System (RAPS): Mathematical Formulations. DOE/RL/87-09, PNL 6200, Department of Energy, Washington, D.C.

Yeh, G. T., and V. S. Tripathi. 1989. A critical evaluation of recent developments in hydrogeochemical transport models of reactive multichemical components. Water Resources Research 25(1), 93–108.

Appendix: Biographical Sketches of Committee Members

FRANK W. SCHWARTZ received a Ph.D. in geology in 1972 from the University of Illinois. He is currently a professor at the Ohio State University; until very recently he was at the University of Alberta. In addition to his research and teaching, he has been an active consultant to government and private industry since 1972. Most of this work has involved project management, report review, technical advice, the development and application of computer models, and field investigations.

CHARLES B. ANDREWS received a Ph.D. in geology in 1978 from the University of Wisconsin, Madison. Since 1984 he has been vice president, corporate office, at S. S. Papadopulos & Associates, Inc., where he directs projects involving all aspects of quantitative ground water hydrology. Areas of expertise include the formulation of ground water projects, modification and development of new off-the-shelf numerical simulation models for adaptation to specific field projects, and evaluation of contaminant and energy transport in ground water systems. Current interests lie in developing techniques for quantifying the risk associated with a given level of contamination in ground water when only limited data are available. Previously, he served with Woodward-Clyde Consultants, Walnut Creek, California, as senior project hydrogeologist.

DAVID L. FREYBERG received a Ph.D. in hydrology, hydraulics, and hydromechanics from Stanford University in 1981. Currently, he is associate professor of civil engineering at Stanford. His research and teaching focus on geologic variability and the quantification and control of uncertainty in ground water transport prediction. Prior to 1980 he was a project engineer and project manager, water resources management department, Anderson-Nichols and Co., Inc., Boston. Since 1985 he has been recognized as a Presidential Young Investigator. Dr. Freyberg is also a member of the Water Science and Technology Board committee currently evaluating the proposed National Water Quality Assessment Pilot Program.

CHARLES T. KINCAID, a senior research engineer in the hydrology section of Battelle's Geosciences Department, received a Ph.D. in engineering in 1979 from Utah State University. He is currently the acting section manager for hydrology and the group leader for soil physics at Battelle, responsible for professional staff who are studying mathematical models of physical processes and chemical reactions, analytical and numerical code developments, and site-specific applications of models/codes. He has specialized in the area of computational fluid mechanics and has experience in both finite-difference and finite-element numerical methods and their application to surface and subsurface flows.

LEONARD F. KONIKOW received a B.A. in geology in 1966 from Hofstra University. His graduate studies were at Pennsylvania State University, where he received an M.S. and a Ph.D. in geology in 1969 and 1973, respectively. He has worked for the Water Resources Division of the U.S. Geological Survey since 1972 and currently is a research hydrologist in their Reston, Virginia, office. He was selected by the Hydrogeology Division of the Geological Society of America to be the Birdsall Distinguished Lecturer for 1985–1986. His work focuses on the development, documentation, and application of solute transport models for ground water contamination problems.

CHESTER R. McKEE is president of In-Situ, Inc., a consulting firm in Laramie, Wyoming. He received a B.S. in physics from Duquesne University and a Ph.D. in hydrology from the New Mexico Institute of Mining and Technology. Prior to forming In-Situ in 1975, he worked at the Lawrence Livermore Laboratory on energy-related subsurface hydraulic problems. For the past 12 years, McKee's consulting experience has included hydrologic evaluations and environmental licensing related largely to mining projects. He has published about 20 articles on hydrodynamics, explosive fracturing, subsidence, hydrology, and restoration of ground water.

DENNIS B. McLAUGHLIN received a Ph.D. in 1985 from Princeton University. His current research interests include the effects of spatial variability and data variability on the accuracy of ground water models. He is associate professor of civil engineering at the Ralph M. Parsons Laboratory, Massachusetts Institute of Technology, having previous experience at the University of California, Davis, as lecturer, and as principal, Resources Management Associates, Lafayette, California. His principal fields of interest are hydrology and water resources systems.

JAMES W. MERCER received a Ph.D. in geology in 1973 from the University of Illinois. He is president and hydrogeologist, GeoTrans, Inc., specializing in all phases of geohydrologic transport analysis, including ground water flow, and heat and solute transport in porous media for a wide range of applications. Previously he was hydrologist, U.S. Geological Survey, Water Resources Division, Reston, Virginia. He has served on a National Research Council panel on ground water contamination, on an advisory panel on national ground water contamination for the Office of Technology Assessment, and on a ground water research subcommittee of the Environmental Protection Agency's Science Advisory Board. He is a member of the National Research Council's Water Science and Technology Board.

ELLEN J. QUINN received an M.S. in management in 1986 from Rensselaer Polytechnic Institute. Since 1982 she has served as scientist, Northeast Utilities Service Company, Hartford, Connecticut, where she designs projects to investigate conditions at hazardous waste sites; evaluates chemical data and water flow information to determine the extent of contamination and recommend remedial measures; and negotiates with regulatory agencies to obtain required permits and approval of site studies. She has developed computer modeling capability for analysis of ground water flow and chemical transport to determine compliance with state and federal regulations. Previously she was a consultant at Sandia National Laboratory, Albuquerque, New Mexico. Prior to that she was project manager, U.S. Nuclear Regulatory Commission, Division of Waste Management, Washington, D.C.

P. SURESH CHANDRA RAO received a Ph.D. in soil physics in 1974 from the University of Hawaii. Currently, he is professor of soil physics at the University of Florida. His research interests are in the development and field testing of process-level models for predicting the environmental fate of pollutants. He is also currently working with state regulatory agencies in Florida on evaluating computer models that can be used to forecast potential ground water contamination from pesticide use. Professor Rao is also a member of the Water Science and Technology Board of the National Research Council.

BRUCE E. RITTMANN is professor of environmental engineering in the Department of Civil Engineering at the University of Illinois at Urbana-Champaign. He received a Ph.D. from Stanford University. His expertise lies in biological approaches to water treatment, including contaminated ground water and aquifers. His research has emphasized the biodegradation of trace concentrations of organic compounds and biofilm kinetics.

DONALD D. RUNNELLS received a Ph.D. in geology from Harvard University in 1964. Currently, he serves as professor of geological sciences at the University of Colorado, Boulder. His research has been in geochemistry of natural waters, low-temperature geochemistry, water pollution, geochemical exploration, and geochemistry of trace substances. He has served previously as geochemist, Shell Development Company, Texas and Florida, and assistant professor of geology, University of California, Santa Barbara.

PAUL K. M. van der HEIJDE received an M.S. at the Technical University at Delft, Netherlands. Currently, he is director of the Water Science Program, International Ground Water Modeling Center, Holcomb Research Institute, Butler University, Indiana. His research has centered on application of ground water hydrology, advancing the use of quality-assured modeling methodologies in the management of ground water resources, and development of the technology transfer methods in ground water science. He is a member of the American Geophysical Union and the Royal Institute of Engineers, the Netherlands.

WILLIAM J. WALSH received a B.S. in physics from Manhattan College and J.D. from George Washington University, Washington, D.C., in 1978. Since August 1986, he has practiced environmental law at Pepper, Hamilton & Scheetz, where he is a partner. Previously, he served as lead attorney for the Environmental Protection Agency in the "Love Canal" and related litigation involving four large hazardous waste landfills in the Niagara Falls, New York area. He is directly familiar with many of the legal and technical issues arising when an attempt is made to utilize contaminant models in regulatory proceedings. He is a member of the New York Academy of Sciences and the American Geophysical Union.

Index

A

D

Q

R

T

U

V

W